Gandhi Meets Primetime

POPULAR CULTURE AND POLITICS IN ASIA PACIFIC

Series Editor
Poshek Fu,
University of Illinois at Urbana-Champaign

A list of books in the series appears at the end of this book.

Gandhi Meets Primetime

GLOBALIZATION AND

NATIONALISM IN

INDIAN TELEVISION

Shanti Kumar

UNIVERSITY OF ILLINOIS PRESS

URBANA AND CHICAGO

1 2 3 4 5 C P 5 4 3 2 1

Library of Congress Cataloging-in-Publication Data

Kumar, Shanti.

Gandhi meets primetime : globalization and nationalism
in Indian television / Shanti Kumar.

p.   cm. — (Popular culture and politics in Asia Pacific)

Includes bibliographical references and index.

ISBN 0-252-03001-X (cloth : alk. paper)

ISBN 0-252-07244-8 (pbk. : alk. paper)

1. Television broadcasting—India.

2. Television broadcasting—Social aspects—India.

I. Title.

II. Series.

PN1992.3.I4K86   2006

384.55'0954—dc22   2004029691

CONTENTS

ACKNOWLEDGMENTS

This book is the result of a long process of research, learning, and unlearning. None of it would have been possible without the generous help and cooperation of my teachers, colleagues, students, friends, and family.

I am deeply grateful to my colleagues in the Department of Communication Arts at the University of Wisconsin–Madison for their unmatched collegiality and steadfast support. I cannot express my heartfelt appreciation to all of them individually here, but I want to especially thank Tino Balio, David Bordwell, Vance Kepley, Steve Lucas, J. J. Murphy, and Sue Zaeske. I hope they know how important their encouragement has been to me over the years.

To my mentors Michael Curtin, Julie D'Acci, and Michele Hilmes, I owe an immense debt of gratitude. They have always been the most astute of my critics and the most gracious of my advocates. I thank them for their warm friendship and exemplary scholarship. Above all, I appreciate their generosity of spirit.

I thank Hemant Shah for being a wonderful friend and colleague. With his characteristic wisdom, wit, and patience, he has supported my work in countless ways. I also wish to thank Andy Sutton and other members of the Media, Performance, and Identity Research Circle for providing intellectual and financial support for my work from a very early stage. I also acknowledge the financial support I received from the International Institute and the Graduate School at the University of Wisconsin–Madison.

I must thank my colleagues at the Center for South Asia, Global Studies, and Borders and Transnational Cultural Studies for being exceptionally supportive of my research interests. My sincere thanks go to Vinay Dharwadker, Sharon Dickson, Joe Elder, Jo Ellen Fair, Susan Friedman, Ken George, Mark Kenoyer, Anne McClintock, Kirin Narayan, Rob Nixon, and V. N. Rao.

Among the many friends and colleagues who have helped me along the way, I would like to thank Bettina Becker, Chris Chekuri, Tim Havens, Olaf Hoerschelmann, Kevin Howley, Radhika Parameswaran, Angie Statz, Atsushi Tajima, Chad Tew, and Serra Tinic. Thanks also to Melissa Curtin for being such a thoughtful friend, and to Lisa Parks for being a generous collaborator.

Many thanks to Chris Anderson, Purnima Bose, and Dwight Brooks. Their constructive criticism and sustained engagement with my work have been invaluable to this book. I also acknowledge, with gratitude, the contributions of Anantha Babbili, Neil Easterbrook, Andy Fort, and David Whillock.

Special thanks to Poshek Fu, my editor at the University of Illinois Press. His creative insights, editorial guidance, and unwavering commitment sustained me through the ups and downs of the writing process. Without his support, this book would never have seen the light of day. Thanks also to Joan Catapano for having faith in this project from the start and to Angela Burton for her timely help at critical junctures in the production process.

I benefited enormously from the insightful comments offered by the anonymous readers of my manuscript. I thank them for their thoughtful reviews. Martha Ramsey has my gratitude for her expert copyediting, and I note my appreciation to Deby Bors and Thomas Vecchio for preparing the index.

Last but not least, I would like to thank my family for their patience, encouragement, and understanding. Madhavi gave generously of her time to help me see this book to completion. With unbounded enthusiasm and good cheer, she provided intellectual insights and emotional support as only she can. For that, and much more, I will always be grateful. P. C. Saan and Vadina are always there for me, and I can never thank them enough for everything they do. Sangita, Sanskrita, and Dharani, you make my research trips to India so much more fun.

To my parents, I owe everything. I would not have been where I am today without their blessings. I dedicate this book to my parents, and to the loving memory of my sister, Janaki.

Gandhi Meets Primetime

Introduction: Unimaginable Communities

WE LIVE IN A DYNAMIC WORLD of electronic capitalism where traditional definitions of nationality, community, and identity are always in flux. We are only beginning to understand the significance of transnational networks such as CNN, MTV, and STAR TV, which can bypass national governments and can connect with television viewers with the click of a remote-control button. We have scarcely recognized the growing influence of translocal media networks such as Eenadu TV, Sun TV, and Zee TV, which can strategically use linguistic appeal to affiliate with the vernacular interests of domestic viewers and diasporic communities. We are witnessing the digital convergence of traditionally distinct telecommunication technologies such as satellites, cable television, telephones, computers, and fiber optics, which has enabled media networks to at once broadcast globally, narrowcast locally, and simulcast for multiple viewership. With the growing trend toward international migration and intercultural mobility, the conventional equation of "communities" with national boundaries is becoming increasingly contentious. The location of culture is being constantly reconfigured on and off television through alternative imaginations of time, space, history, geography, identity, and difference.

In this book, I argue that the rapid transformations of electronic capitalism in general and the growing competition among television networks in particular have necessitated radical reimaginations of nationalism in postcolonial India. I frame the problematic of nationalist imagination in Indian television in relation to the cultural thematics of identity and difference

outlined by Partha Chatterjee in his critique of Benedict Anderson's formulation of nations as "imagined communities."[1] According to Chatterjee, the
cultural elites in colonial India imagined the nation into existence not in an
identity with but rather in a difference from the British colonizers. However, in the postcolonial context, I argue that the colonial distinctions of
print-capitalism—such as the colonized and the colonizer, inside and outside,
us and them—have been blurred by the rapid growth of electronic capitalism, and a new generation of media elites have mobilized television to articulate (i.e., link) hybrid imaginations of identity and difference to idealized
notions of Indian nationalism.

Even a cursory glance at the changing landscape of Indian television
reveals that the meteoric rise of satellite and cable channels in the 1990s has
disrupted the hegemony of the state-sponsored network, Doordarshan, in
unparalleled ways. Although government monopoly still dominates terrestrial broadcasting through the airwaves in India, the emergence of foreign
and domestic networks since 1991 brings into focus a series of very important
questions about the role that television plays in articulating diverse imaginations of the nation as a community. When viewers across the country tuned
in to *Ramayan* and *Mahabharat,* television productions of the great epics,
in the late 1980s, it was perhaps possible to argue that Doordarshan had a
hegemonic hold over representations of nationalism in Indian television.[2]
However, even in the pre–satellite television era, Doordarshan always had to
contend with the popularity of commercial cinema and the vernacular diversity of print media. Therefore, I argue that the recent influx of satellite
and cable channels has further contributed to the vigorous contestation of
Doordarshan's hegemonic status as the national network in India. In this
heterogeneous terrain of the mass media, I seek to critically interrogate diverse imaginations of nationalist identity and cultural differences in broadcasting, satellite, and cable television in India.

In the first section of this introduction, I discuss how the emergence of
transnational networks, such as STAR TV, and translocal networks, such as
Zee TV, Sun TV, and ETV, has led to the creation of hybrid electronic vernaculars that seek to cater to the globalist aspirations, the nationalist inspi-
rations, and the regionalist affiliations of the media elites in India. In the
second section, I outline a theoretical framework to address the changing
relationship between electronic capitalism and postcolonial nationalism in
Indian television. In the third section, I describe the overall objectives of this
study and summarize the rest of the book.

STAR on the Horizon

In January 1991, when an international alliance of forces led by the United States attacked Iraq in response to Saddam Hussein's invasion of Kuwait in August 1990, the American cable news network CNN was on scene in Baghdad to telecast the war live all around the world. Media scholars contend that the Persian Gulf war was "perhaps the first war ever orchestrated for television."[3] CNN's presentation of the war was highly dramatized for effect on television, and viewers across the world could hear "exclamations of 'oohs' and 'oow'" as loud explosions from bombing interrupted the CNN correspondents' live reports in Baghdad.[4] In India, some hotels, businesses, and high-rise apartments in major cities like Mumbai and Delhi that had dish antennae to receive satellite channels tuned in to CNN's live coverage of the war from Baghdad.

Many five-star hotels that had satellite dishes on their rooftops found a sudden increase in calls for room reservations. Hotel Maurya in New Delhi even imposed a cover charge of 100 rupees for access to their bar, which provided customers with access to satellite television. Hotel Asrani International in Hyderabad reportedly had to hire an additional twelve rooms a day to accommodate customers who wanted to watch the Gulf War on the hotel's cable television system. After watching CNN, a viewer complained that Doordarshan's coverage of the Gulf War was "like a Scud missile that failed to explode—a real damp squib."[5]

Since many Indians have friends and family members working in the Persian Gulf region, concerned viewers in these cities were even willing to pay satellite dish owners to watch CNN, which was their only source for twenty-four-hour news about the war in Iraq. Cat Vision, a satellite dish manufacturer in Delhi, which also claimed the franchise for CNN coverage in India, reported a threefold increase in the demand for its dish antennae the day after war broke out in the Persian Gulf. Electrocom Systems, an Ahmedabad-based dish antenna dealer, reportedly ran out of stock, as requests poured in from other dealers in towns like Rajkot and Surat, and even smaller places in the region like Junagadh, Dhoraji, and Upleta.[6]

After the Gulf War ended in February 1991, the intense interest in CNN slowly began to wane among those who had tuned into its Gulf War coverage for over a month. However, in May 1991, a little-known network called Satellite Television Asian Region, or STAR TV, started broadcasting over

Asia from Hong Kong, using a leased satellite called *Asiasat*. Initially, there were four channels on the STAR TV platform: Star Plus, an entertainment channel largely made up of Euro-American programming; Star Sports, once again a channel with largely Euro-American sporting events; BBC News, the British Broadcasting Corporation's worldwide news service; and MTV, the American music channel's joint venture with the Asian satellite network.

Some satellite television dealers, such as Shyam Communication Systems in New Delhi, bought advertisements in leading national newspapers and magazines and offered to set up a private TV reception system for anyone interested in having six channels of STAR TV at their fingertips. The cost of an eight-inch dish antenna was listed at 25,000 rupees, with additional charges for installation and taxes. Claiming to be India's number one satellite systems network, Shyam Communications sent out the following invitation to interested viewers: "Now plug in to the excitement of 6 world-class channels, simultaneously, 24 hours a day with STAR TV, MTV, BBC, Prime Sports, Family Entertainment and two international language channels. Flip the channels and choose what you want to see, not what the cable TV operator decides you should see" (see figure 1).

Soon, a number of enterprising cable operators—or *cablewallahs,* as they came to be called—hastily assembled satellite dishes in their backyards and offered to connect homes in their middle-class neighborhoods for a modest monthly fee of 60–100 rupees (approx. U.S.$2–3). In less than one year after the first dish antenna was realigned toward CNN, a modest satellite and cable industry was born, and television viewers suddenly found themselves exposed to a variety of hitherto-unheard voices and unspoken messages.[7]

In stark contrast to the censored news, regulated documentaries, patriotic songs, and nationalist sitcoms on the state-sponsored network, Doordarshan, STAR TV seemed to offer viewers a varied and novel fare: old and new American soaps like *Dynasty* and *The Bold and the Beautiful,* titillating talk shows like *Ricki Lake* and *Oprah Winfrey,* dramatic shows like *Baywatch,* live coverage of international sports, MTV, and BBC World Service. Very quickly, STAR TV became a rage among the affluent sections of the English-speaking Indian middle classes, and the nascent satellite and cable television industry in India began to witness radical transformations.

According to a study conducted by the marketing research agency Frank Small, in the first six months of 1992, cable operators in India were hooking almost 4,500 homes up every day. That figure rose sharply to 9,450 homes daily in the second half of the year. From a modest figure of 412,000 urban

Figure 1. Advertisement for Shyam Communication System, phone: 91-11-25798544; FAX: 91-11-25790762. *India Today,* February 29, 1992, 4.

households in January 1992, the cable television industry expanded to 1.2 million in November 1992, and 3.3 million households in January 1994. In another study conducted at the end of 1994, Frank Small estimated that out of a total of 32.4 million homes with television, the number of cable television households in India was 11.8 million.[8]

In 1995, national estimates provided by the Indian Readership Survey suggested that there were about 8.4 million urban homes and 3.4 million rural homes with cable television in India. However, the National Readership Survey conducted in the same year estimated that the number of urban homes with cable television was 9.3 million. At the same time, other surveys commissioned by key players in the satellite and cable industry were more optimistic, placing the national figure at about fourteen million households. By the end of the 1990s, the Indian television industry estimated that there were at least twenty-two million cable television households in a national population of over seventy million homes with television.[9]

These national figures, when multiplied by an average of five individuals in each household, suggest that there are at least 110 million viewers with access to satellite and cable television in India and over 350 million viewers who can tune into the state-sponsored network, Doordarshan, which has the potential to reach 87.6 percent of the population, and can cover 72.9 percent of the national community in terms of its geographic area.[10]

In July 1993, Rupert Murdoch's News Corporation bought 63.6 percent of the company's total shares from the then owners of STAR TV, Hutchinson Whampoa and the Li-Ka-Sheng family in Hong Kong. In July 1995, STAR TV came to be wholly owned by Rupert Murdoch when News Corporation purchased the remaining 36.4 percent of the shares. Recognizing the need to compete with other major networks for a potentially vast but currently limited viewership in Asia, STAR TV embarked on a "think global, program local" policy.[11] The strategic mission of STAR TV's "glocalization" policy has been twofold: (1) to provide a more "localized" perspective to the network's visibly global image, and (2) to embrace a programming strategy that appeals to the nationalist sentiments of their diverse audiences.

However, Rupert Murdoch's STAR TV soon recognized that programming for an elite pan-Asian viewership posed a number of problems. First of all, the geographical dispersion of television viewers across Asia made programming far more complicated, given the differences in time zones. For example, programming for late night in Taipei might not be acceptable during primetime in Mumbai. Second, these dispersed Asian viewers were difficult to monitor. Viewership studies and television ratings were almost nonexistent during the early 1990s, and the prospect of mounting pan-Asian surveys seemed particularly daunting for STAR TV. Yet without such studies, it was difficult to set advertising rates and even more difficult to convince sponsors to buy commercial time. Finally, the number of corporations that were interested in transnational advertising was not as large as originally imagined. Network executives began to realize that they needed a mix of transnational, national, and even local advertising if they were to meet their revenue objectives.[12]

During the 1990s, almost every transnational satellite television network worth a name—CNN, the Cartoon Network, ESPN, Disney, and Discovery Channel—set up shop in Asia. To reach the geographically dispersed and culturally diverse Asian viewers, some networks distributed their services through satellite transmission, while others entered into strategic partnerships with state-owned terrestrial broadcasters.

In 1996, a Goldman Sachs report revealed that in the vast Asian markets, STAR TV had quickly emerged as the largest transnational television network, with an estimated potential of fifty-three million households. These estimates, reported by Goldman Sachs, included forty-seven million households in sixteen countries capable of receiving STAR's service, and an additional 5.85 million households in market areas not surveyed.[13] Although STAR TV now covers over fifty countries in Asia and the Middle East, the Goldman Sachs report outlined an uncertain future for the pan-Asian network during the 1990s. Despite the market uncertainty and financial setbacks in the beginning, Murdoch made it clear to journalists and media analysts that "STAR's value to News Corp . . . is its building asset value."[14]

In 1992, News Corporation acquired a 49.9 percent share in a Hindi-language channel called Zee TV, launched by an Indian businessman, Subhash Chandra. When Chandra joined the STAR TV bandwagon and started telecasting soaps, sitcoms, and talk shows on Zee TV, Indian audiences were for the first time exposed to Hindi-language television that was not Doordarshan, the state-sponsored national television network. Chandra's strategy was to provide the Hindi television viewers with what Doordarshan did not; a Hindi-language equivalent of STAR TV's English-language fare of soaps, sitcoms, talk shows, game shows, and the like.[15]

Moreover, unlike the state-sponsored Doordarshan, commercial networks like Zee TV were not constrained by direct government control over programming content. This factor alone was an adequate incentive for Indian viewers to watch political debates on television without bureaucratic interference by government officials. A popular weekly program on Zee TV was *Aap Ki Adalat*, in which politicians, celebrities, and business tycoons were openly accused of corruption and interrogated in a rapid-fire question-and-answer session by a real-life lawyer named Rajat Sharma. To account for the popularity of commercial networks like Zee TV in India, Manjunath Pendakur and Jyotsna Kapur identify a number of attributes of shows like *Aap Ki Adalat* that viewers find riveting:

They [the viewers] want to see the politicians who are key to certain national crisis pinned against the wall. Because political power is the domain
of the rich, powerful, and/or criminal elements in society and because it
touches their lives so deeply, the audience love to see the politicians un-
masked, albeit briefly. Even if they wiggle out, their sweaty foreheads, embarrassment, awkward grinning, when they seem to [be] caught by the
clever questioning and evidence presented by the attorney, are entertaining.
The audience identifies completely with the host, who is smart, informed,
well dressed, and clearly appears to be on the people's side. The laugh track
ruptures the false idea created by the show that it is a court of law, but
nobody seems to care.[16]

In order to recognize the political appeal of the visibly simulated debates
on shows like *Aap Ki Adalat,* Indian viewers only had to compare it with
similar programming (or lack thereof) on the state-sponsored network. As
Pendakur and Kapur remind us, if a show like *Aap Ki Adalat* had been shown
on Doordarshan, "there would have been political hindrance placed in its
way by the Prime Minister's office or someone in the Ministry of Information and Broadcasting."[17] Although the blatantly political process of deci-
sion-making at Doordarshan always hindered programming on the state-
sponsored network, the commercial networks like Zee TV had their own
hidden economic agendas that could interfere with the creative act of production. To put it plainly, in airing controversial shows like *Aap Ki Adalat,*
commercial networks like Zee TV were less concerned with Doordarshan's
political objectives of creating and sustaining a unified sense of nationalism
and more interested in exploiting the economic incentive that television
provides by generating "sufficient numbers of those desirable audience members with disposable income to spend on the goods that are advertised on
the show."[18]

Nevertheless, the popularity of shows like *Aap Ki Adalat* fleetingly revealed a subtle yet significant distinction between the programming strategies of Doordarshan on the one hand and commercial satellite networks
such as Zee TV on the other. In their eagerness to gain dominance in the
economic sphere, media elites at the helm of commercial networks like Zee
TV revealed an almost suicidal tendency to disrupt the hegemony of the
ruling classes in the political sphere. This was a political-economic trade-off
that the government officials at Doordarshan had always been unable or
unwilling to make in their attempts to create a unified "national programming" schedule for Indian television.

The popularity of Zee TV's commercial programming enabled the network to break even in the first year of its operations. By the mid-1990s, Zee TV established itself as a major network catering to Hindi-speaking viewers, with an estimated reach of 140 million people in twenty-five million households in India, and an additional seventy million audiences in fifteen million households in other markets around the world. In 1995, Zee TV established services in the United Kingdom. In 1996, it extended its network into Africa, and by 1998, it entered the lucrative satellite television market in the United States.[19] In September 1999, Chandra further consolidated his media empire by buying out Murdoch's stake in Zee TV. The political significance of this seemingly minor economic exchange in the global media industry was not lost on media analysts in India. As Prasun Sonwalker argues, when the STAR–Zee TV partnership began in 1993, "few could have foreseen that Chandra, not Murdoch, would emerge stronger and richer out of the deal."[20]

In the wake of the rising popularity of Zee TV's programming strategies in India, other privately owned, commercial networks, such as Sony TV, began offering varied Hindi-language programming. On its part, STAR TV found it necessary to further localize its services by transforming itself into a hybrid "Hinglish" network, programming in Hindi and English at once. In 1997, Rajat Sharma of *Aap Ki Adalat* fame was contracted by STAR TV to produce a program called *Janata Ki Adalat* in an obvious attempt to compete with the growing success of Zee TV's programming formats in Hindi.[21]

Other commercial networks, such as Sun TV and Eenadu TV, started programming services in regional languages, especially in South India, where English and Hindi are spoken only by a limited few in the large metropolitan areas. Catering to the vernacular interests of television viewers, regional-language networks steadily grew in the 1990s by wedging their claims to public legitimacy into the cultural flux generated by the success of STAR TV and Zee TV. The strategy of the regional-language channels was to create a translocal television network that catered to viewers' interests in ways that the centralized authority of Doordarshan never did appropriately and the English-language STAR TV and the Hindi-language Zee TV could not ever do adequately.

The emergence of Eenadu Television in the 1990s as a translocal network catering to a variety of regional-language interests in India is a case in point. Eenadu Television is part of the Eenadu Group, a large, regional media conglomerate located in Hyderabad, the capital city of Andhra Pradesh. Although a relative newcomer to the television scene, the Eenadu Group has been in the media business for over three decades and has diverse interests, ranging from financial services, shipping, and pickles and spices to advertising, publishing, and film production and distribution. *Eenadu* newspaper, the leading Telugu daily in Andhra Pradesh, has a virtual stranglehold over both advertising and readership in the vigorously competitive print market in the state.

One of the most successful newspapers in the vernacular in postcolonial India, *Eenadu* was founded by Cherukuri Ramoji Rao, a Telugu entrepreneur who began his media career in advertising. When he launched the *Eenadu* in 1974, it had a modest beginning as a single-edition, hand-composed newspaper that was printed on a flatbed press in the costal town of Visakhapatnam in Andhra Pradesh. From the beginning, Ramoji Rao intensely researched the Telugu newspaper industry and used creative marketing strategies to sell *Eenadu* to both advertisers and readers alike. By the early 1990s, the Eenadu Group had transformed into a major regional conglomerate, with daily editions of the newspaper from seven towns, which boasted a combined circulation of almost 550,000.[22]

When Eenadu Television—or ETV as it is popularly known—was launched in August 1995, there was great anticipation among Telugu-speaking audiences who were already familiar with the parent company's powerful influence in Telugu media, thanks largely to the success of the daily newspaper by the same name. Leasing a high-quality transponder on the Intelsat satellite system, and uplinked from Padduka near Colombo, Sri Lanka, Eenadu began television transmission with an ambitious seventeen-and-a-half-hour service of entertainment and film-based programming. The day after the inauguration of ETV, Indrajit Lahiri of the *Times of India* reported on the network's programming mix as follows.

> Children's programmes include *Shabash Tintin* and *Baboi Dennis* [the English series *Adventures of Tintin* and *Dennis the Menace* dubbed into Telugu], *Master Minds*, a children's quiz and *Andhravani*, a daily news capsule for children. Among its film-based programs are *Film Masala*, *Studio Roundup*, *Cinema Quiz*, *Swa Gataalu* in which veteran stars take viewers down memory lane. Another film-based show is *Hats Off*, a mix of film songs, dances and scenes, and *Sarigamalu*, an *Antakshari* . . . started by famous music director S. P. Balasubramanium. There are also programmes for women like *Sneha*, *Vasundhara*, *Chitti Chitkalu*. *Ghuma Ghumala*, a fifteen-minute show on household tips by famous stars and also featuring two of their favorite recipes, is a novel idea. Eenadu also offers programmes for the family like *Countdown*, *Comic Tonic*.[23]

When Gemini TV was launched in April 1995, Telugu-speaking viewers had access to two privately owned channels in the state of Andhra Pradesh.

Owned by the Gemini Group, which runs film production facilities, film-processing laboratories, and hospitals in South India, Gemini TV began with a modest four-hour service via a Rimsat satellite uplinked from Singapore. Early programming was mostly film based—drawing from the Gemini studio's archives of song-and-dance sequences from popular Telugu cinema. In 1996, Gemini TV was bought by the Sun TV network, which is owned by Kalanithi Maran, the media-savvy son of India's minister of commerce and industry, Murasoli Maran, and the grandnephew of the chief minister of Tamil Nadu, M. Karunanidhi. In 1999, Maran's Sun network—consisting of Sun TV in Tamil, Udaya TV in Kannada, Gemini TV in Telugu, and Surya TV in Malayalam—captured over 50 percent of the advertising revenue in the South India television market.

A cover story in *Business World* on Maran's growing media empire estimated that Sun TV had cornered over 2,500 million rupees (approx. $56 million) of the total (approx. $100 million) advertising revenue of 4,500 million rupees for the Indian channels (including Doordarshan).[24] Television advertising in South India is a small but rapidly growing segment of the overall advertising pie in India, which was estimated to be around 25,000 million rupees (approx. $556 million) in the year 2000.

When Maran's Sun TV took over Gemini, they quickly recognized that neither the initial euphoria of its launch nor its impressive library of song-and-dance sequences from Telugu films were enough to sustain viewers' attention in the crowded satellite television market. To attract the Telugu-speaking viewers, Gemini quickly expanded its programming schedule and began commissioning independent producers to create serials, sitcoms, soap operas, and current affairs and film-based shows. In addition, it started dubbing popular American television shows and Hindi television epics like *Ramayan* and *Mahabharat* into Telugu.

To consolidate its position as one of the early leaders in the Telugu television market, Gemini began courting advertisers and sponsors with brand-name programs like *Santoor Top Ten, Thumbs Up My Choice,* and *BPL Sarigamalu,* to name a few. By the mid-1990s, thirteen out of the thirty-nine programs aired weekly on Gemini Television were brand named. Outlining the network's marketing strategy, Abbi Karan of Gemini TV remarked: "Through these programs we are analysing the likes and dislikes of viewers which will enable us to develop more successful software later."

According to market surveys conducted before the launch of ETV and Gemini TV in 1994, there were about 4.5 million television sets in Andhra Pradesh and about 1.3 million cable households. While Gemini TV boasted a reach of 85 percent of cable households, ETV was so confident about its potential to conquer the Telugu cable market that the channel did not even conduct a survey before launching its service.[25] The confidence derives in no uncertain measure from an awareness of the viewers' familiarity with the "Eenadu" brand image that the television channel proudly proclaims as its unique selling point. Calling it the "ETV" advantage for its Telugu viewers, an ETV Publicity Report, *Tune In,* released in 1996 proclaims: "Today the Telugu viewer has no real choice of programmes, in his own mother tongue. Entertainment-starved, he has settled for badly produced programmes or satellite channels in English or Hindi. So, if you were to give him quality programmes and in his own mother tongue, rest assured he would tune in. And tune in to ETV—a quality family entertainment channel in Telugu" (see figure 2).

The obvious embellishments of publicity and promotion notwithstanding, the ETV report points to the perennial wisdom of the Telugu viewers in no uncertain terms: that is, all things considered, language and cultural location are central to the understanding of television. In this context, *eenadu*—a polysemic Telugu term that means "this day" and "this land"—not only represents the brand image of a major regional media conglomerate in Hyderabad but also articulates the creative ways in which translocal media networks have created an electronic vernacular using the linguistic appeal of cultural proximity and regional identity to compete with their national and transnational counterparts. While Eenadu—this day—promotes the news value of timeliness for instantaneous communication of public images in the vernacular, Eenadu—this land—emphasizes the news value of proximity for intimate communion with private imaginations that enhance the everyday experience of Telugu audiences.

Elated by the audience's response to its Telugu language fare in Andhra Pradesh, Eenadu earnestly expanded its network into other Indian states and offered viewers across the country programming in the regional language of their choice. Similarly, Sun TV and Zee TV rapidly expanded their networks in the 1990s and have offerings in languages ranging from Tamil, Telugu, Kannada, and Malayalam to Marathi, Punjabi, Gujarati, and Bengali, to name a few.

The rise of "Hinglish"-language programming on transnational networks like STAR TV, and the subsequent growth of regional-language programming by translocal networks like Eenadu TV, Sun TV, and Zee TV, demonstrates how the medium of television has engendered a hybrid category of "electronic vernaculars" in postcolonial India. Occupying a strategic location

Tune In

1. A unique entertainment channel that is all set to reach over seven crore Telugus. 2. Transponder on Intelsat IS-704 for quality reproduction of signals right from day one... 3. A true family channel with entertainment for all age groups. 4. A satellite channel from the Eenadu Group.

Figure 2. Publicity report for Eanadu Television, touting its "family entertainment channel." *Tune In,* Eenadu Television Report, 1996, 1.

between the national language of Hindi and the transnational language of English, the electronic vernaculars of the translocal networks, such as ETV, Sun TV, and Zee TV, are sufficiently localized to cater to the linguistic affinities of regional communities, and are adequately standardized to provide a global marketplace to advertisers and sponsors who seek to attract domestic and diasporic viewers to their brand of commercial products, ranging from soaps and cereals to cars and computers. However, the economic incentives for media networks like ETV, Sun TV, and Zee TV to articulate their versions of the electronic vernacular are not predefined by traditional imaginations of nationalism or limited by the political boundaries of the nation-state in postcolonial India. Sevanti Ninan makes this point well in the following observation: "When a Ramoji Rao or a Subhash Chandra assesses the market for a Bengali channel, he takes into account both the market in West Bengal as well as in Bangladesh and other parts of the world where Bengalis are. . . . In other words, the multi-channel scenario at home is made viable by the globalisation of regional entertainment."[26]

As the economic incentives of a multichannel marketplace drive media elites such as Kalanithi Maran, Ramoji Rao, and Subhash Chandra to reimagine Indian television both within and beyond the political boundaries of the nation-state, they lead to the creation of what I call the *unimaginable communities* of electronic capitalism.

Imagined Communities, Real Differences

The formulation "unimaginable communities" is an obvious allusion to Benedict Anderson's influential work *Imagined Communities*. However, embedded in that allusion is also the call for a critical revision of Anderson's insights into the role of imagination in the origin and spread of nationalism in the colonial and the postcolonial worlds.

According to Anderson, the nation must be understood as an *imagined* community "because the members of even the smallest nation will never know most of their fellow-members, meet them, or even hear of them, yet in the minds of each lives the image of their communion."[27] The nation is imagined as a *limited* community, he continues, "because even the largest of them, encompassing perhaps a billion living human beings, has finite, if elastic boundaries, beyond which lie other nations."[28] Finally, Anderson declares, the nation is imagined as a community "because regardless of the actual inequality and exploitation that may prevail in each, the nation is always conceived as a deep, horizontal comradeship."[29]

Unlike the imagined communities of print-capitalism that Anderson defines as being finite, limited, and bounded, the unimaginable communities of electronic capitalism I posit are infinite, limitless, and unbounded in the worldwide flows of national, transnational, and translocal networks. The electronic flows of television are *unimaginable* both in a literal sense of being at the technological limits of imaginative access and in a figurative sense of becoming limitless in imaginary excess. The electronic flows of television engender *communities* both in a sacred sense of communion and a secular sense of communication, in that the everyday transmissions of television channels attain the meaningful status of images only in an intimate encounter with the diverse imaginations of viewers at home.

To posit a decisive break between the unimaginable communities of electronic capitalism and the more traditionally imagined communities of print-capitalism is not to fetishize the power of television as a radically new technology. Rather, it is a relational argument about the dual role of television in simultaneously enabling and disrupting the process of imagining nations as communities in the world of electronic capitalism.

I posit the notion of unimaginable communities to describe the many ways in which people who may never meet each other physically can come, quite literally, face to face on television. In this sense, the impact of electronic capitalism on the television viewer's sense of space and place is more dynamic—and therefore more difficult to imagine—than the conventional formulations of nationality and community in print-capitalism. For one thing, the electronic mobility engendered by television enables viewers to be part of the secular community of nationalism, or the divine communion of religious traditions, even as it opens the possibilities for transgressing the arbitrary boundaries that a nation-state must enforce to protect its territorial sovereignty, or a religion must impose to protect its transcendental authority. In what follows, I elaborate on the paradoxical role of electronic mobility in the unimaginable communities of Indian nationalism by critically evaluating the crucial role that television plays in articulating the many uses and abuses of Mahatma Gandhi's status as the Father of the Nation of postcolonial India.

Sivanti Ninan speculates that if the Spirit of Mahatma Gandhi was watching over the spate of programming invoking his name on Indian television, "it would have been startled to discover how much in vogue he still is."[30]

Sampling the television coverage on Gandhi's birthday on October 2, Ninan finds that an MTV veejay, Nikil Chinnappa, wished a caller "Happy Gandhi Jayanti!" While a pony-tailed anchor on Kairali hosted a Gandhi memorial quiz with the Indian flag prominently displayed in the background, the internet was used by some to call on Indians in the United States to "observe global vigil for peace." Not to be outdone, the transnational channel STAR News carried pithy quotes from Mahatma Gandhi's extensive writings on the principles of nonviolence. "What took the cake," Ninan writes, was the following comment by a news anchor on Doordarshan's *Business Day:* "Today is Mahatma Gandhi's birthday that is why the markets are closed, it is also Lal Bahadur Shastri's birthday, and it is also the birthday of my father whom I wish as a proud son. Happy Birthday."[31]

Although Ninan invokes the Spirit of the Mahatma to survey the simultaneous exposure and erasure of Gandhian nationalism on Indian television, it is perhaps more appropriate in this context to speak of the specters of Gandhi rather than of a Spirit gazing from afar. The notion of such a spirit conjures up a transcendental image of a whole and a holy subject/object that is merely reflected without any distortion by the material and ideological apparatuses of television. However, what we see of Gandhian nationalism in India is always partial—in the two senses of the word—and therefore a spectral image that is sometimes a subject, at other times an object, and at many times neither.

Neither visible nor invisible, neither soul nor body, neither an icon nor an idol, neither a phantasm nor a simulacrum, neither living nor dead, the specter is, as Jacques Derrida explains, "both one and the other."[32] Mediating the limits of the visible and the invisible, the specter of Gandhi on Indian television is an unimaginable, or almost unimaginable, "Thing" that "looks at us and sees us not see it even when it is there."[33] To begin with, there is the specter of Gandhi the man, who was born Mohandas Karamchand Gandhi.[34] Then, there are many specters of "Bapu," the Father of the Nation, so hailed for his leadership during the Indian nationalist struggle against British colonialism. There are the state-sponsored specters of Gandhi as "Mahatma" in countless hagiographies, documentaries, news programs, and national integration messages on Doordarshan. There are commercially sponsored specters of Gandhi too, shown in ads to sell everything from sewing machines and ethnic-chic T-shirts to Bajaj automobiles and Macintosh computers.

There is, of course, the Hollywood Gandhi, neatly captured by Richard Attenborough's Oscar-winning spectacular *Gandhi,* produced in 1983. There are many specters of Gandhi in Bollywood too, Kamal Hasan's *Hey Ram* and

Shyam Benegal's *The Making of the Mahatma* being two of the more recent ventures in a long list of big-budget and small-budget films of this kind produced in India. Often, there are independent films, documentaries, dramas, and plays that seek to bring nuance—even controversy—to Gandhi's role as the Father of the Nation; Anand Patwardhan's *War and Peace,* Vinay Apte's *Mee Nathuram Boltoy,* Chandrakant Kulkarni's *Gandhi Virudh Gandhi,* and Centan Datar's *Gandhi-Ambedkar,* to name a few. There is also a primetime Gandhi on satellite and cable television channels—a spectral figure that is at once tragic, comic, and heroic in its cameo appearances on late-night talk shows, MTV top-ten rotations, twenty-four-hour newscasts, soap operas, and sitcoms.

Despite a plethora of plays, documentaries, TV serials, Hindi films, and even Hollywood blockbusters that have been produced over the years, Mukesh Khosla finds that Mahatma Gandhi "never fails to inspire writers, directors, actors and viewers" who seek to unravel "the Mahatma behind Gandhi."[35] Among the many efforts is a weekly serial on Doordarshan entitled "Meri Kahani" (My Life). Based on Gandhi's writings on his own life and drawing from Louis Fischer's insightful biography *The Life of Mahatma Gandhi,* the serial begins in 1921 with an image of Gandhi writing his autobiography. Gandhi is presented as the *sutradhaar* (narrator) of his life story through the nationalist struggle against British colonialism, and the serial cuts back and forth between dramatic re-creations of events shot on location and old stock footage from historical archives of documentaries in the Films Division of India.

Yet another serial on Doordarshan, entitled *Mahatma,* attempts to capture little-known details about the life of Gandhi as a child and a young boy in the 1870s and 1880s. This fifty-two-episode teleserial is based on Gandhi's autobiography, *My Experiments with Truth.* The narrative is supplemented by other historical resources gathered from the Gandhian archives at the Navjivan Trust Library in Ahemdabad.[36]

On April 26, 2001, when STAR Plus launched *Ji Mantriji*—the Indian version of the BBC serial *Yes Minister*—a portrait of Mahatma Gandhi was prominently displayed behind the minister's desk. The irony of it was not lost on commentators reviewing the show's opening episode. As one reviewer put it, "the decision to film in Hindi reveals how ingrained colonial traditions are in India, a country that borrowed, then refined the British bureaucratic model to the point where it produced concepts such as the VVIP—Very, Very Important Person."[37] Of course, even the most important of these VVIPs in India must pay homage to and be blessed by Mahatma Gandhi in his exalted role as the Father of the Nation. The prominent display of Gandhi's portrait in the backdrop of *Ji Mantriji*'s contemporary political setting is a subtle reminder of how the BBC serial has been carefully adapted to suit the Indian context in which Gandhian nationalism is a silent but ever-present specter.

Not to be left behind, regional-language networks have also invoked Gandhian ideals to connect with the nationalist sentiments of their vernacular viewers. The Asiasat network commissioned a well-known Malayalam filmmaker, Viji Thampi, to produce a serial comedy called *Mahatma Gandhi Colony*. Thampi explains: "*Mahatma Gandhi Colony* is an attempt at portraying a sort of mini India within the bounds of a residential colony."[38] Although a comedy serial, the television show also aspires to represent the Gandhian ideal of Indian nationalism by portraying characters who, as Thampi puts it, are "from every walk of life with different natures and different ways of life etc., and who live in harmony with each other."[39]

In spite of—or perhaps because of—the prevalence of television shows, newspaper headlines, dramatic productions, and films that hail Gandhian nationalism in postcolonial India, controversy seems to always surround Gandhi's name and image, whether one encounters them in mundane discussions about of the virtues of vegetarianism and the vices of uninhibited consumption or in profound debates about the merits and demerits of *swaraj* (self-rule) and *swadeshi* (self-reliance) in national governance and international relations.

In 1992, the producers of *Humrahi*, a television serial on Doordarshan, were startled by instructions from officials at the state-sponsored network to delete certain remarks made by film actress Tanuja, who performed the role of the anchor of the show. The offensive remarks in question included a line about how "persecution of man by man goes on in the land of Mahatma Gandhi." Taken aback by the criticism from Doordarshan officials, the producers of the show had no other choice but to delete the offensive remarks before airing the episode on April 28, 1992.[40] A furor erupted over a fashion show held in the city of Jaipur when a model wore a costume that had on the back a picture of Mahatma Gandhi sitting at a handloom wheel spinning cotton. Since this image was widely used during India's freedom struggle against British colonialism as a symbol of nationalist independence, many Gandhians were outraged and filed a case in the public interest at the Jaipur High Court. While registering the case, the presiding judge ordered that the organizers of the fashion show apologize to the nation for the inappropriate display of Gandhi's image on a model's back.[41]

Another major controversy ensued after an inappropriate remark about

Mahatma Gandhi was uttered on the episode of *Nikki Tonight* that was aired on the STAR TV network on May 4, 1995. In this ill-fated episode of the titillating talk show hosted by Nikki Bedi, Ashok Row Kavi, a prominent gay rights activist and journalist in India, called Mahatma Gandhi "a bastard *bania.*" *Bania* is a Hindi word used, often pejoratively, to refer to the community of traders from which Mahatma Gandhi hailed in the northwestern Indian state of Gujarat. Tushar Gandhi, a great-grandson of the Mahatma, filed a lawsuit and appealed to the national government to protect the legacy of his family. As furious politicians across party lines in Parliament raised concerns about cultural imperialism through satellite television, some parliamentarians even demanded that government ban transnational networks like STAR TV from operating in India.

Sensing the intensity of the public ire in response to the controversy, the producers of *Nikki Tonight,* Nikki Bedi, and Row Kavi all issued apologies. A few days later, STAR TV quietly withdrew the talk show and acknowledged that the network had erred in not editing the program before airing it. STAR TV executives also sought to debunk theories of cultural imperialism by noting that the show was produced by TV-18, an Indian company based in Mumbai and the views expressed in the show were those of an activist-journalist in India, and therefore were not indicative of the transnational network's regard for the nation's revered figures such as Mahatma Gandhi.

Despite such assertions by the media elites, there is a great deal of apprehension in some sections of Indian society that transnational networks such as STAR TV are new incarnations of the East India Company, which colonized India under the pretext of creating trading outposts for the British empire in the eighteenth century. Therefore, controversial episodes involving the name of Mahatma Gandhi are never about, as Sidharth Bhatia puts it, "silly television programs or naive hostesses" on Indian television.[42] Rather, the controversies surrounding Gandhi's status as the Father of the Nation are indicative of "the sensitive and delicate balance between India and the West" that television networks have to constantly mediate in their efforts to create innovative programs and profitable schedules.[43]

If I draw attention to the specters of Gandhi haunting Indian television, it is not to construct a master narrative of postcolonial nationalism in the name of its venerated patriarch. Rather, I analyze the scattered references to Gandhi's name and image as a way to interrogate the role that television plays in the articulation of nationalism to electronic capitalism in postcolonial India. I draw upon a range of materials, including contemporary television programming, historical archives, legal documents, policy statements, academic writings, and journalistic accounts. The empirical evidence is illuminated by theoretical analyses that combine diverse approaches, such as cultural studies, poststructuralism, and postcolonial criticism.

I deconstruct images in print advertisements from the early 1970s to the end of the 1990s to trace the changing status of the television set as a cultural commodity in postcolonial India. I also analyze publicity brochures, promotional materials, and programming schedules of translocal networks such as Eenadu Television to outline the role of vernacular media in the discourse of electronic capitalism. I closely examine a variety of television texts, such as national integration messages on Doordarshan and controversial talk shows like *Nikki Tonight* on STAR TV. Each of the following chapters invites the reader to ask the question "Is there an Indian community of television?"—each time in a different way—by focusing on the various roles that Mahatma Gandhi plays in the articulation of nationalism to electronic capitalism.

Is There an Indian Community of Television?

In a charmingly informal essay entitled "Is There an Indian Way of Thinking?" A. K. Ramanujan invites the reader to engage in a Stanislavsky-like exercise by asking the question four different ways, placing emphasis on different parts of the question each time.[44] Extending Ramanujan's dramatic exercises in theatrical performances to the realm of electronic media, I add a fifth question to focus attention on the role of *television* in mediating collective imaginations of nationalism in postcolonial India. Therefore, I ask the question "Is there an Indian community of television?" in five different ways:

> *Is* there an Indian community of television?
> Is there *an* Indian community of television?
> Is there an *Indian* community of television?
> Is there an Indian *community* of television?
> Is there an Indian community of *television*?

Depending on where one places the emphasis, the question contains many other questions that, as Ramanujan reminds us, are "are real questions—asked again and again when people talk about India"[45] The answers, as Ramanujan reveals, can be just as many. All of us who have any pretensions to some kind of engagement with India—academic or otherwise—have given one or many of these answers at different times to different questions about the national community of Indian television. "I, certainly . . . have," writes Ramanujan, recalling the various answers to questions about an Indian way of thinking.[46] I confess that I too have held different views at different times when responding to different questions about Indian television.

In chapter 1, I trace the evolution of Doordarshan as a state-sponsored network in relation to debates over the creation of an autonomous Prasar Bharati Corporation to oversee broadcasting in India. In this changing context of electronic capitalism, I answer the question "*Is* there an Indian community of television?" by defining the dissemination of nationalist autonomy in Doordarshan and Parsar Bharati in terms of a hybrid formulation of imagiNation.

In chapter 2, I deconstruct diverse representations of television as a cultural commodity through close textual analysis of print advertisements that appeared in leading national news magazines such as the *Illustrated Weekly of India* and *India Today* from the early 1970s to the end of the 1990s. I contend that the question of *an* Indian community of television can be addressed in terms of a bifocal vision that enables the viewer to be at home in the world of electronic capitalism.

In chapter 3, I focus on the theories of national development proposed by Indian leaders such as Gandhi who sought to articulate an uncolonized vision of *swaraj* (self-rule) and *swadeshi* (self-reliance) as a postcolonial alternative to both Western capitalism and Soviet communism. In this overall framework of international relations, I conclude that the question of an *Indian* community of television must be understood in terms of a nationalist struggle to imagine an "uncolonized" community based on Gandhian principles of independence such as *swaraj* and *swadeshi*.

In chapter 4, I address the question of how an Indian *community* of television is articulated in the name of Gandhian nationalism in a newspaper article written by a prominent Indian feminist scholar, Urvashi Butalia (1994). While Butalia's article delves into the problems and potential of imagining the nation as a community after the arrival of satellite television channels in 1991, the headline "Gandhi Meet Pepsi" aptly describes how Gandhi's name is symbolically overwritten to death by the competing discourses of nationalism and electronic capitalism in Indian television.

In chapter 5, I focus on the fifth question, "Is there an Indian community of *television?*" By analyzing the aforementioned *Nikki Tonight* controversy on STAR TV, I foreground the central role that television has come to occupy in mediating diverse imaginations of nationalism in India by at once deifying and defiling the exalted status of nationalist icons like Mahatma Gandhi. I argue that *Nikki Tonight* is an exemplar text that reveals how televisual imaginations of national identity and cultural differences have become ever more important, even as they are increasingly being blurred by the dynamic flows of electronic capitalism in postcolonial India.

In the final chapter, I conclude this study by revisiting the five questions about the Indian community of television and summarizing the findings of each chapter. By focusing attention on the question "Is there an Indian community of television?" in five different ways, I contend that television has become the new battleground for competing visions of postcolonial nationalism by fueling fears of foreign domination through satellite channels, even as it takes the national community further into the world of electronic capitalism.

1 From Doordarshan to Prasar Bharati: The Search for Autonomy in Indian Television

AROUND 3:00 P.M. on November 12, 2001, the viewers of Doordarshan and the listeners of All India Radio were treated to a rare address that had been broadcast to the nation by Mahatma Gandhi on the same day in 1947. The historic event was recreated to commemorate the fifty-fourth anniversary of Gandhi's first and last visit to the Broadcasting House in Delhi to record his message to refugees who had been violently displaced from their homes by the Partition of India and Pakistan. At the commemoration ceremony, some of Gandhi's favorite *bhajans* (devotional songs) were also played, and a new museum of radio and television was inaugurated at the Broadcasting House by the minister of information and broadcasting, Sushma Swaraj.

Describing the museum as the first of its kind in the country, Swaraj declared that the historic event was a reminder of the need for "introspection by the broadcasting agency to contemplate upon its goals as a public service agency." Anil Baijal, the CEO of the Prasar Bharati Corporation, which oversees both radio and television broadcasting in India, explained that the goal of this endeavor was "to preserve our heritage and present it in the shape of a museum for the future generations." Baijal also reminded the audiences that in the previous year, Prasar Bharati had declared November 12 "national broadcasting day" in honor of Mahatma Gandhi's historic address to the nation in 1947.[1]

In this chapter, I interrogate the nationalist quest for autonomy in public broadcasting in India by mapping the changing identity of Doordarshan as a state-sponsored network. In the first section, I trace the historical evolution of Indian television from the early 1950s to the late 1970s, when the nationalist agenda of educational programming defined the identity of Doordarshan as a state-sponsored network. In the second section, I examine how the rapid commercialization of Doordarshan through the 1980s contributed to the creation of "national programming" as a genre that articulated the state-sponsored agenda of educational television to consuming desires of the cultural elites for entertainment. In the third section, I discuss how the growing challenge from satellite and cable television channels in the 1990s induced the government of India to recast Doordarshan's identity as a public broadcasting service through the creation of an autonomous corporation called Prasar Bharati. In the final section, I address the question of an Indian community of television in terms of a hybrid formulation of imagiNation as a way to articulate the nationalist ideals of public broadcasting to the rapid transformations of electronic capitalism.

Creating a National Network

The idea of television broadcasting in India was first suggested in October 1951 by a Scientific Advisory Committee that was set up by the government to explore the possibility of establishing a pilot station. On February 2, 1953, the union minister of information and broadcasting announced plans to establish an experimental project to examine whether television would be within the economic means of the Indian government. Commenting on the feasibility of such a project, the minister declared: "Though television might appear to be a useful thing in the country, the expenses involved in installing it are very high."[2] India's first prime minister, Jawaharlal Nehru, was very hesitant to commit the government's limited resources for the very high expenses necessary to sustain the electronic medium of television. However, major players in the Indian electronics industry proposed to explore the possibility of commercial operations by organizing public demonstrations of television in major cities like Delhi and Bombay.

In 1959 Philips India set up a demonstration of closed-circuit television at the Industrial Exhibition in Delhi. At the conclusion of the exhibition, Phillips sold the broadcasting transmitter and twenty-one television sets for a fraction of their cost to the government of India. Subsequently, a UNESCO grant of $20,000 enabled the government of India to purchase fifty-five additional television sets, which were set up for community viewing in and around Delhi. A pilot broadcasting center was set up in the Delhi premises of All India Radio (AIR), and a small team of producers and engineers began experimenting with educational programming and the technical evaluation of broadcasting equipment.[3]

On September 15, 1959, experimental television services were inaugurated in Delhi by the president of India, Dr. Rajendra Prasad. The experimental service was limited in scope and had specific objectives: to create television programming of educational and cultural value to both urban and rural communities. As part of this pioneering experiment, sixty-six "Tele-Clubs" were organized in adult education centers in and around Delhi to receive adult education programs that were broadcast for one hour twice a week on Tuesdays and Fridays. The Tele-Club organizers were trained by All India Radio researchers to conduct follow-up discussions after the programs and to keep a report of the viewers' reactions for later evaluation.[4]

In January 1960, All India Radio, in collaboration with the Delhi Directorate of Education, began producing one-hour educational programs for students in higher secondary schools. Since access to television sets was very limited, viewing was organized in Tele-Clubs under the guidance of "teacher-sponsors" who were responsible for presenting the programming material to the students in ways that were relevant to the prescribed school syllabi. Subjects covered under the educational television program included science, history, health and hygiene, language training in Hindi, and community affairs. Sometimes student-produced puppet shows and plays were included in the one-hour program.[5]

To further explore the feasibility of educational television in India, the Ford Foundation sent a team of experts from the United States to visit India in January 1960. This team examined some of the educational television programs made by the All India Radio staff, visited a number of schools in Delhi, and conducted several interviews with schoolteachers, principals, and administrators in the Delhi Directorate of Education. On the basis of their recommendations, the Ford Foundation granted $564,000 to the government of India as partial support for a four-year educational project using 250 television sets in nearly three hundred higher secondary schools with more than 150,000 students in Delhi.[6]

In December 1964, the government of India appointed a committee, headed by Asok K. Chanda, to evaluate the performance of, and recommend appropriate changes to, the various media units under the Ministry of Information and Broadcasting. In its report, the Chanda committee proposed the separation of television from the organizational structure of All India

Radio and recommended the creation of the autonomous Television Corporation of India to avoid political interference from government officials in the Ministry of Information and Broadcasting.[7]

On August 15, 1965—the eighteenth anniversary of Indian national independence—general television services were launched with daily, one-hour transmissions from Delhi. Although entertainment and informational programming was introduced as part of the "General Service," the proclaimed goal of television broadcasting in India was educational, and programming emphasized issues such as adult literacy and rural development. General Service consisted of a ten-minute "News Round Up," mostly read by an on-screen presenter in a format developed for audio broadcasts on All India Radio.[8]

Visual relief was provided by broadcasting the "Indian News Review," produced by the Films Division of the government of India, which supplied a number of documentary films for television. In addition, free documentaries were also available from the embassies of foreign governments eager to project a positive image of their country abroad. The film-based programs were converted into television format using a small sixteen-millimeter telecine unit—once again, a gift of the United States Information Service. Since the telecine unit did not have provision for thirty-five-millimeter films, telecast of film-based materials, including documentaries and feature films, was initially restricted to the sixteen-millimeter format. Needless to say, the technical quality of the sixteen-millimeter films after conversion into television format left much to be desired. Nonetheless, given the novelty of the medium, neither the producers nor the viewers seemed to complain too much about the blemishes in broadcasts.[9]

On January 26, 1967—Republic Day—Prime Minister Indira Gandhi inaugurated *Krishi Darshan,* a pilot project aimed at evaluating the role of television in the mass dissemination of educational and informational programming for rural development. Cosponsored by the Indian Agricultural Research Institute, the Department of Atomic Energy, the Ministry of Information and Broadcasting, All India Radio, and the Delhi administration, the *Krishi Darshan* program catered to rural viewers, who were provided with community television sets in eighty villages near Delhi. With the active involvement of the government of India and its various agencies to promote broadcasting as a medium for national development, television left its Delhi moorings, as transmission centers were set up in cities and towns across India during the early 1970s.

On October 2, 1970—the birthday of Mahatma Gandhi—a television production and transmission center was set up in Mumbai (then Bombay). A cosmopolitan city with a diverse group of viewers in terms of language, religion, class, caste, and communities, Mumbai represented a challenge for broadcasting, since programming from Delhi was restricted to the national language, Hindi. Although attempts were made to produce programming in Marathi, the major language spoken in the state of Maharashtra, Mumbai had a large number of viewers who understood very little of English, Hindi, or Marathi, since they spoke other languages such as Gujarati, Konkani, Urdu, and Tamil, to name a few.

On Republic Day, January 26, 1973, a television center was established in the city of Srinagar, as the government felt it was necessary to have broadcasting services in the state of Jammu and Kashmir in order to counter programming flowing into India from across the border in Pakistan. Concerns about national security also contributed to the launch of the next television center in the city of Amritsar on September 29, 1973. With programming in English, Hindi, Punjabi, and Urdu, the Amritsar center catered to viewers in the Indian state of Punjab but also reached across the border into the city of Lahore in the Pakistani province of Punjab. Indian cinema has always been very popular in Pakistan, and when the Amritsar center broadcast film-based programming, viewers in and around Lahore flocked to their television sets, much to the consternation of the government of Pakistan. At the same time, television serials produced in Pakistan were very popular among Indian viewers, who had very little interest in tuning into the state-sponsored propaganda that was transmitted from the Amritsar center in the form of news and documentaries.[10]

To counter the popularity of Pakistani television serials, the government of India mobilized the production centers in Delhi and Amritsar to provide entertainment programming for viewers in the Indian state of Punjab. However, in these early years, the broadcasting centers in Amritsar and Delhi did not have the production facilities, personnel, or resources to provide a constant flow of entertainment programming to counter Pakistani serials from across the border. Therefore, Indian films and film-based programming in Hindi and Punjabi became a major component of entertainment programming at the Amritsar center.[11]

In June 1975, when Prime Minister Indira Gandhi declared a state of national emergency in response to growing opposition to the authoritarian policies of her government, severe restrictions were placed on the freedom of press and political assembly in the country. At the same time, the state-sponsored media of radio and television were used for political propaganda about Indira Gandhi's "Twenty-Point Programme," in an effort to enlist the nation's support for her government's policies of national development during the Emergency. The prime minister's son, Sanjay Gandhi, who had no official role in the government, used television to promote his own "Five-Point Programme" and mobilized overzealous sycophants in the Congress Party to implement his pet projects of forcible evacuation of slum-dwellers in Delhi and compulsory family planning for the poor and the illiterate.[12]

To ensure the mass dissemination of the state-sponsored propaganda during the Emergency, the Indian government continued to spread its television network into major cities and towns across India. When the Calcutta television center was started on August 8, 1975, it presented a peculiar set of challenges to the centralized structure of broadcasting in India. The capital city of West Bengal, Kolkata (then Calcutta) was—and still is—a stronghold of Marxist and communist parties that wielded enormous political influence in the state with little or no opposition from the Congress Party. Needless to say, the political aspirations of the Marxists and communists in the state were never adequately represented in the national network, since it was centralized in its organization and controlled from Delhi by bureaucrats and politicians in Indira Gandhi's Congress Party government.

When the Madras center was launched on August 15, 1975—commemorating the day of national independence—it brought into relief another dimension of the tensions between the states and the central government. The capital of the south Indian state of Tamil Nadu, Chennai (then Madras), was—and still is—the epicenter of the DMK (Dravida Munetra Kazagam, or the Dravidian Progressive Party) and the AIADMK (All India Anna Dravida Munetra Kazagam, or the All-India Dravidian Progressive Party started by Anna Durai) parties, which have effectively mobilized the linguistic affinities and cultural identities of the Tamil community in Tamil Nadu to fight against the centralized authority of the Congress Party at the national level.

During the early years of radio broadcasting in postcolonial India, political leaders in Tamil Nadu—even those in the Congress Party—had been vociferous in their demands for state-level autonomy and often criticized the government of India for marginalizing regional languages and for promoting Hindi as the national language. A major controversy over language erupted in 1957, when the Directorate General of Information and Broadcasting issued a memorandum conveying the government's desire to change the name of All India Radio to "Akashvani" (*akash*, "skies"; *vani*, "voice"). Like all stations in the nation, the Tiruchi station of All India Radio in Tamil Nadu implemented the new policy, but the station director sent the Directorate General a copy of the Tamil newspaper *Dina Thomthi,* which reported on an agitation against the "imposition of Hindi." H. R. Luthra recounts the ensuing controversy in the days following the incident:

> The agitation mounted despite clarifications made by AIR that only where the reference was to "All India Radio" was the word, Akashvani to be used, but where the reference was to Radio in general local equivalents like "Vano-li" or "Nabhovani" may be used. Dr. Keskar, Minister for I & B [Information and Broadcasting] wrote to Mr. Bhaktavatsalam, Home Minister, Madras[,] on August 25, 1958[,] that All India Radio had adopted "Akashvani" as its all-India name in India, and "it had now become a kind of trademark." He added that the word Akashvani had been taken from Kannada (where it was the name originally given to the Mysore station from the British days onwards), and it had been adopted because it was easily understood everywhere in the country. While, therefore, there was no objection to "Vanoli" it was not possible for the Radio to drop Akashvani, which had been adopted as a patent word for the service.[13]

Although the chief minister of Tamil Nadu, Kamraj Nadar, assured the government of India of the state's cooperation in implementing the new policy on All India Radio, some Tamil groups continued their agitation into May 1959, with individual agitators threatening to fast unto death if the name change was not withdrawn. At the heart of this contentious struggle over nomenclature is a set of articles (343–351) in the Constitution that prescribe Hindi in the Devanagari script as "the official language of the Union" with the continued use of English as "a subsidiary national language" as long as is necessary to facilitate communication among the various agencies of the national and state-level governments in the Union.[14]

Article 351 of the Constitution provides for "the promotion and development of the Hindi language so that it may serve as a medium of expression for all the elements of the composite culture of India." To ensure the composite expression of India's diverse linguist communities, the Constitution directs all Indians to "secure the enrichment" of Hindi as a national language "by assimilating without interfering with its genius, the forms, style and expressions used in Hindustani and in other languages of India specified in the Eighth Schedule, and by drawing, wherever necessary and desirable for its vocabulary, primarily on Sanskrit."[15]

On April 1, 1976, when television was uncoupled from radio through the official institution of the national network, Doordarshan, it was evident that the constitutional directive to use Hindi by drawing primarily from Sanskrit was still being followed in letter and spirit by the government of India, which seemed to have forgotten earlier protests over the Sanskritization of All India Radio to *Akashvani*. The use of the prefix "door," meaning tele or distant, and *darshan,* meaning vision or sight, and the introduction of the term "doordarshan" to describe television and the corresponding term "darshak" to refer to viewers are attributed to J. C. Mathur, who was the director general of All India Radio when experimental television services were launched in 1959.[16]

To sustain the state-sponsored agenda of national development, programming on Doordarshan had an almost exclusive focus on issues like agriculture, animal husbandry, poultry farming, education, literacy, health, and family welfare. However, other programming genres, like talk shows, quiz shows, children's programs, feature film-based music programs, and sports programs, supplemented the developmental agenda of Doordarshan. Yet, with the goals of national development clearly taking precedence in Indian television, not surprisingly, there was little emphasis on promoting Doordarshan either as a commercial medium for entertainment or as a public enterprise free from government control.[17]

In the general elections of 1977, when the fledgling Janata Party swept into power by defeating the Congress Party, which had suffered its worst electoral losses because of the disastrous policies of the Emergency, there was a major shift in the political climate of the nation. The Janata Party, which contested the elections on a slogan of being the people's party, promised to use its sweeping mandate to make several changes to the centralized media policies of the Congress Party, which had been at the helm of national affairs for three decades since the country had gained its independence from British rule in 1947.

Among the changes proposed by the Janata Party was the appointment of a former newspaper editor, B. G. Verghese, to head an independent commission to suggest ways to decrease government control over broadcasting in India. Following the recommendations of the Verghese committee report, the Janata Party government recommended the creation of an autonomous corporation for public broadcasting, to be called "Akash Bharati" (*akash,* "the skies"; *Bharati,* of India, in Hindi).[18]

While the proposals of the Verghese committee represented a promising start on paper, the political realities of governance soon caught up with the Janata Party, which was being pulled in different directions by the competing interests of its coalition partners. The fall of the Janata Party government and the reemergence of Indira Gandhi's Congress Party in the general elections of 1980 signaled a new phase in the history of Indian television. For one thing, Prime Minister Indira Gandhi's desire to maintain the centralized control over radio and television was well known, and the Janata Party's much-discussed proposal to provide autonomy for the electronic media was immediately shelved by the new government. At the same time, in creating a new ministerial cabinet, Indira Gandhi assigned the influential Ministry of Information and Broadcasting to Vasant Sathe, whose enthusiasm for rapidly transforming Indian television was not always shared by the prime minister or by the other members of the Congress Party.

When New Delhi was awarded the opportunity to host the Ninth Asian Games in 1982, the information and broadcasting minister found himself at the helm of a major overhaul of Indian television in order to introduce color transmission services. Officials at Doordarshan, led by Director General Shailendra Shankar, were elated by the opportunity to revamp the technical infrastructure of the national network, which had been strapped for funds for over a decade. Shankar, who headed Doordarshan from 1980 to 1985, enthusiastically proclaimed: "When the history of Indian television is written, Sathe's name should be in colour."[19] Thanks to Doordarshan's coverage of the Asian Games, the relatively small but significantly powerful middle class in India, which had little to associate with the early developmental agenda of television, had their first taste of continuous entertainment programming—that too in color.

The Nationalist Colors of Indian Television

Following the introduction of color transmissions on August 15, 1982, and the successful coverage of the Asian Games in November of the same year, officials at Doordarshan began working toward a massive expansion of the national network. In December 1982, the Ministry of Information and Broadcasting set up a Working Group under the chairmanship of P. C. Joshi "to prepare a software plan for Doordarshan, taking into consideration the main objectives of television of assisting in the process of social and economic development of the country and to act as an effective medium for providing information, education and entertainment."[20] In its two-volume report, the Joshi committee pointed to the fundamental importance of "software planning" in the process of creating an "Indian Personality for Television."[21]

While the recommendations of the Joshi committee emphasized the need for software planning in the creation of an "Indian" personality for television, officials at the Ministry of Information and Broadcasting focused on hardware development to rapidly expand the geographic reach of the national network. In 1983, potential coverage for Doordarshan grew from 23 percent to 70 percent of the population, as the number of television transmitters increased from 41 to 180 in less than one year. S. S. Gill, the secretary of the Information and Broadcasting Ministry, proudly announced that the country had created "the biggest information explosion in the history of communication."[22]

As part of this ambitious plan for the expansion of Indian television, high-power transmitters and low-power relay stations were set up in major cities and small towns across India. The national network was decentralized only to the extent that large metropolitan cities like Mumbai, Bangalore, Kolkata, Chennai, Hyderabad, and a few other smaller cities had regional-language transmission centers. Each of these centers catered to the specific regional interests of their viewers, who all shared a common regional language, usually within a single state. Although the languages were different in each of the centers, the programming formats and content remained largely similar across the national network. The political bosses in Delhi managed to sustain their vision of an Indian community of television through a loosely defined yet extremely centralized authority over the administration and programming of the national network.

On July 7, 1984, Doordarshan began broadcasting *Hum Log* (We the People), a partly educational and partly entertainment television serial that was based on a communication strategy that was designed by Miguel Sabido in Mexico to produce "telenovelas" (soap operas) to encourage social change and national development. As Arvind Singhal and Everett M. Rogers argue, "*Hum Log* was an attempt to blend Doordarshan's stated objectives of providing entertainment to its audience, while promoting, within the limits of a dominant patriarchal system, such educational issues as family planning, equal status for women, and family harmony."[23]

The story of *Hum Log* revolves around the everyday activities of a North Indian joint family, with each episode focusing on the triumphs and tribulations of one or more of the nine central characters, who span three generations. Although everyday conflicts and tensions in relationships between parents and children, grandparents and grandchildren, siblings and cousins provided the necessary elements to serialize the episodic narrative, nationalist issues of patriotic pride, family planning, gender relations, and communal harmony also became central to the definition of *Hum Log* as the story of an "Indian" family. The nationalist imagination in *Hum Log* is furthered by the appearance of Ashok Kumar, the well-known Hindi film actor, at the end of each episode to summarize the central elements of the twenty-two-minute narrative.

In his commentary, Ashok Kumar—lovingly called Dadda as a mark of respect for his status as a grandfatherly figure (or an elder brother in Bengali)—would address the narrative conflicts and moral dilemmas of *Hum Log* in terms of their relevance for a typical Indian family that is caught between the cultural tensions of tradition and modernity in everyday life. At the end of his one-minute monologue, Ashok Kumar would leave the story in tantalizing suspense, with an invitation to the viewer to find out what happened by tuning in to the next episode of *Hum Log*. Over the course of the 156 episodes of *Hum Log* in 1984–85, as Ashok Kumar signed off each episode by translating the show's title into a different Indian language, his one-minute summary came to symbolize Doordarshan's programming agenda of creating a collective union of the nation's diverse linguistic and ethnic communities.

At the same time, the broadcasting of *Hum Log* also signaled a new era of commercialization on Indian television, as Doordarshan entered into a contract with Food Specialities Limited, the Indian subsidiary of Nestle, to sponsor the production of the serial. Under this arrangement, Food Specialities agreed to pay for the production costs of *Hum Log* in return for the rights to nationally advertise its product—Maggi Two-Minute Noodles—during, before, and after each episode, which, in 1984–85, reached about sixty million viewers across India. After Food Specialities began its sponsorship of *Hum Log*, there was a dramatic increase in the national sales figures for Maggi noodles. The sales for Maggi noodles "increased from none in 1982 to 1,600 tons in 1983, 4,200 tons in 1985, 10,000 tons in 1990 and 15,000 tons in 1998."[24] Soon, the Food Specialities campaign for Maggi Noodles was considered a textbook example of how a company could advertise and market new consumer products on Indian television.

The enthusiastic response of Indian viewers to entertainment programs like *Hum Log*, and to new consumer products like Maggi Two-Minute Noodles, prompted advertisers to compete for commercial sponsorship on Doordarshan, which then had a monopoly on television viewing in India. Arvind Rajagopal describes the ambivalent relationship, in this monopolistic framework, between the government's agenda of national integration on Doordarshan and the commercial incentives for private corporations to sponsor programming on television as follows.

As notions of entertainment culture (as well as their more concrete embodiments by way of imported technology and software) "leak" and spread into more countries, private entrepreneurs enter the business, in films, video and cable (satellite, of course, represents a new challenge, the terms of engagement with which are still emerging). The state may turn to entertainment culture as follower rather than initiator, on the heels of a growing entrepreneur's market. State-owned media may then be used as a way of seeking control over a growing popular culture serving an influential sector of the public, and potentially threatening in its autonomy. Broadcasting, unlike other public utilities, needs to create varied rather than unvarying products to fulfil its function. Being highly rule-bound and structured, however, state-controlled media are often unable to generate the innovative capacity required to attract audiences. State sponsorship of private production thus becomes an attractive option. Audiences then begin to move towards television as a fixed-cost investment, and advertising money follows audiences.[25]

With commercialization of national programming on the rise by the mid-1980s, an emerging generation of postcolonial elites quietly discarded the early designs for using Doordarshan as a public medium for national development in favor of a more entertainment-oriented commercial culture. Led by Rajiv Gandhi—now the prime minister of the nation—a younger, more urban, Anglophile, and technophile generation took charge of producing a new, more cosmopolitan image for Indian television. In this changing scenario, as Ananda Mitra points out, the noneducational programming genres in Indian television began to diversify "from the earlier dependence on feature-film based programs into television plays . . . serials and soap operas."[26] For instance, in the "What's on TV" section of the March 1987 issue of *TV and Video World*, Mitra finds eleven different serials in Hindi and English in the national programming schedule of Doordarshan. These include, Mitra writes,

> *Sara Jahan Hamara,* a program that tells the story of "13 lovable, generous and impressible 'brats' who set about tackling life on their own terms in an orphanage"; *Khoj,* a story about a lady detective; *Mickey and Donald,* the cartoon series; *Kashmakash,* which features 13 short stories all by Indian women writers; *Swayam Siddha,* which traces the story of [the transformation of] a woman from a "vulnerable, confused and unsure person" [in]to a confident woman; *Subah,* which engages in the debates surrounding the quality of contemporary college life in India; *Ek Kahani,* about a set of villagers who struggle with oppressive land owners; *Buniyad,* the so-called Indian *Dallas; That's Cricket,* which is a commentary on cricket in India and abroad; *Contact,* a quiz program for school children, and *Ramayan,* the religious soap opera.[27]

As advertising and commercial sponsorship of sporting events, sitcoms, soap operas, dharmic serials, and film-based programs brought in considerable revenues, Doordarshan effectively manipulated its monopoly over viewers across the country by strategically scheduling what the network called national programming during the primetime hours of late evenings and weekends. As Rajagopal defines it, Doordarshan's *national programming* refers to "an emergent category of software in Indian television drawing upon mythological and historical sources, and portraying an idealized past above and beyond latter day divisions."[28] Elaborating on the definition, Rajagopal explains:

> I choose the term "national" to indicate the broad cross-regional appeal of the programmes, and their (usually implicit but sometimes explicit) elaboration of a national culture. The state's appeal to myth and history (intermingled, as always) is instrumental in this purpose. A shared past, behind and above all latter-day divisions, is projected as the crucible in which a distinctive Indian identity was shaped. This identity is, of course, under fierce dispute as competing interests vie to redefine its character; currently, minorities, especially Muslims, are threatened by a blatantly "Hinduized" national identity.[29]

As Rajagopal rightly argues, the emergence of Doordarshan's national programming as a pan-Indian genre was crucial for the postcolonial project of nation-building. To sustain the political agenda of transcending the diversities of language, religion, region, ethnicity, class, caste, and gender in the modern nation-state, the political elites were so drawn to the genre of national programming in television that they were willing to overlook the increasing commercialization of what had been heralded as a public medium at its inception. At the same time, national programming was extremely attractive to the economic elites—ranging from transnational corporations to local businesses and advertisers—whose primary objective was to reach the largest number of television viewers in the cheapest possible way.

In solely economic terms, the political expediency of national programming was irrelevant to advertisers and sponsors who were willing to negotiate with the state-sponsored network in order to gain access to its monopoly over viewership. Thus, national programming on Doordarshan represents a postcolonial genre of Indian television that makes possible the mutual legitimation of the hegemony of the political and economic elites in the national community. In this hegemonic context, the unprecedented popularity of Doordarshan's national programming in the 1980s cannot be understood merely in terms of the political-economic calculations of the media elites. Rather, the reasons for Doordarshan's telecasts of Hindu epics such as *Ramayan* and *Mahabharat* must be examined in relation to the electoral calculations that influenced the Congress Party government to enthusiastically embrace what Rajagopal has described as the "aura of spiritual sanctity" underlying their description as "dharmic serials."[30] Rajagopal writes:

> "Dharmic" in this context refers to matters religious or spiritual, and "serial" is of course a neologism, referring to a periodical issue, in this case of a weekly television program lasting anywhere from thirteen weeks (the typical length of a Doordarshan serial) to two years or more. As a *dharmic* form, the Ramayan serial drew from and appealed to long-standing traditions of attendance at religious story-tellings, *kathas,* which could draw daily audiences running into the thousands for months together.[31]

Drawing on interviews with people who watched *Ramayan* on Doordarshan, Rajagopal describes the ways in which the dharmic serial engendered ambivalent responses to religious traditions and modern nationalism among many Indian viewers. One Sunday morning during the telecast of *Ramayan,* Rajagopal writes, he encountered a group of passers-by gathered around a black-and-white television set in a store. Although one could barely see anything on the screen in the glare of daylight, he recalls one of the viewers saying that people watched the serial out of a sense of devotion because they felt that "God was giving them *darshan.*"[32] The notion of *darshan,* he argues, helps explain how and why Doordarshan's telecasts of *Ramayan* enabled large numbers of viewers who did not share any single political ideology to identify with the images and themes of its weekly telecasts. In defining the congregational experience of *darshan* on Doordarshan as "Prime Time Religion," Rajagopal argues that the television version of *Ramayan* "reformulated the epic in a revivalist manner, projecting a proto-national security state back into the vedic past."[33]

When the Congress Party government had officially sanctioned the broadcast of *Ramayan,* it had hoped to use its religious appeal for the more secular necessities of gaining votes in the elections by attracting a broad base of Hindu viewers into its nationalist fold. However, it did not anticipate that a marginal opposition party, the Bharatiya Janata Party (BJP), would reap the benefits instead. During the course of the *Ramayan* telecasts from 1987 to 1989, the BJP cleverly manipulated the public adulation of the serial to draw attention to a longstanding dispute over the site of a sixteenth-century mosque in

Ayodhya called the *Babri Masjid,* which is also claimed to be the birthplace of the Hindu god Ram—the lead character in *Ramayan.*[34]

In the short term, the BJP and its Hindu nationalist allies, such as the Rashtriya Svyamsevak Sangh (RSS) and the Vishwa Hindu Parishad (VHP), effectively used the popularity of the *Ramayan* serial to draw the attention of television viewers to their demand for the construction of a temple of Ram at the disputed mosque site in Ayodhya. In the long run, however, the BJP was politically invested in using the popularity of the serial to revive its marginalized and much-maligned brand of cultural nationalism called Hindutva, which derives from definitions of India as an exclusively Hindu civilization in origin. For a long time, the Hindutva movement has sought to promote its alternative vision of cultural nationalism as a substitute for the more secular definition of the nation as a multireligious community that is by common consent the state-sponsored ideology in postcolonial India.

Historically, the Congress Party has proclaimed itself to be a secular party and has flaunted its credentials as the protector of religious minorities in India in order to acquire their votes en masse. In the 1980s, the minority vote bank of the Congress Party was fragmented by the rising influence of other national parties like the Janata Dal, the communist parities, and many regional parties, all with pretensions to the influential mantle of secularism. In desperate bids for votes, all national and regional parties have been glaringly guilty of playing a double-sided game by pitching the interests of one religious community against the other to suit their electoral interests.

Cultural nationalists in the Hindutva movement have been very effective in mobilizing *Ramayan*'s popularity to make an emotional appeal to conservative Hindu viewers who perceive the existence of a mosque on Ram's birthplace as a threat to the very foundations of their religion. But what is even more disturbing is the way the BJP and its allies willingly perpetuate this fallacious view of religion and nationalism in their electoral politics. As documented by Anand Patwardhan in films such as *Ram Ke Nam* and *Father, Son and the Holy War,* the advocates of Hindutva nationalism cleverly use antiminority and antiforeign sentiments to narrate a linear history of India that begins with an ancient (native) Hindu civilization that was corrupted for centuries by (external) influences of Islamic despotism and British colonialism.

The essentialist narratives used by the Hindutva forces work in tandem with television serials such as *Ramayan* to legitimize the myth of a precolonial, pre-Islamic subject in the image of Lord Ram as the ideal *native* of the land, and his Hindu descendants as the only ones with an undisputed claim to citizenship in the postcolonial context. The antiminority and antiforeign sentiments also derive their political legitimacy from a small but very influential segment of the media elites who have turned—albeit uneasily—toward the BJP in their disenchantment with the Congress Party's manipulation of the ideals of secularism in the name of nationalist leaders like Jawaharlal Nehru and Mahatma Gandhi.

Thus, the contentious struggle between the forces of Hindutva and secularism has generated a powerful momentum that has the potential to appropriate alternative imaginations of nationalism—based on gender, class, caste, religion, region, language, and so on—into its hegemonic fold. In this struggle, the distinctions between the religious ideals of Hinduism and the political ideologies of nationalism are blurred both on and off screen, and sometimes quite literally so, as in the case of the television series *Mahabharat.*

When it was first broadcast by Doordarshan, in installments between 1989 and 1990, the viewership for *Mahabharat* outscored even the astronomical figures attained by *Ramayan,* which it had replaced in the "primetime religion" hour on Sunday mornings. *Mahabharat* was reportedly watched with ritual regularity in over 90 percent of all Indian television homes, transcending boundaries of religion, caste, class, language, region, and political allegiance. As in the case of *Ramayan,* weekly household routines were reportedly organized around the Sunday telecast, and family TV sets often became the site of community viewing.

The *Mahabharat* story is so diverse and complicated by many tellings in folk theater and literature that the producers of the television version use a figure no less than Time itself to make sense of the many narrative threads of the epic. At the beginning of every episode of the television *Mahabharat,* Time is anthropomorphized as the figure of a sage who sits against a cosmic backdrop of stars and planets, on which is superimposed a slowly turning wheel that signifies the eternal movements of the universe. In the first episode of the *Mahabharat* series, Time cautions against reading *Mahabharat* merely as the tale of a battle between the warring cousins called the Kauravas and the Pandavas.

Claiming to have been a witness to the beginning of the story, Time speaks of an era when the great king Bharat, the ancestor of the Kauravas and the Pandavas, ruled over a vast empire stretching from the oceans in the south to the Himalayas in the north. Time thus invites the viewer to take a journey further into to the past to find the true origins of the great Indian epic. Time also reminds the viewers that the name of the modern Indian nation-state in Hindi, Bharatvarsh, is derived from this originary kingdom of Bharat in *Mahabharat.* In subsequent episodes of the series, Time appears at the beginning to introduce the great diversity of plots, characters, settings and social values in the narrative to help the viewer connect the present with the past. In doing so, I argue, Time also participates in the ideological struggle between the forces of Hindutva and secularism in their efforts to claim both Hinduism and nationalism through the equation of Bharat with India.

Of particular significance to my analysis here is a major segment of *Mahabharat* that is generally referred to as the "Game of Dice." As in the various tellings of *Mahabharat,* this segment of the narrative in the television series begins roughly at the point where tensions between the two clans rise and violent conflicts seem inevitable. The Pandava prince Yudhistra, celebrated as Dharma Raja for his unswerving adherence to dharmic principles, willingly gives up his claim to the throne and relocates his clan to a barren part of the Bharata kingdom in order to avert a bloody family feud. But Duryodhana, the belligerent crown prince of the Kauravas, who envies Yudhistra's nobility and courage, refuses to accept the Pandavas' offer for peace. Aided by his cunning uncle Sakuni, Duryodhana invites the Pandavas to a game of dice, knowing Yudhistra's dharmic penchant for gifting his wealth away. Sakuni uses his evil genius to load the dice in favor of Duryodhana, while the unsuspecting Yudhistra, true to form, gets carried away by the game and loses everything—his wealth, his kingdom, his four brothers, and then himself.

In the climactic scene of the "Game of Dice" episode, Yudhistra takes a final gamble by betting his wife, Draupadi, and ends up losing her to the Kauravas. Intoxicated by his victory, and still intent upon humiliating his defeated cousins, Duryodhana orders his Kaurava brothers to drag Draupadi out of her chambers to be disrobed in front of her five husbands (she is married to all the Pandava brothers). As her husbands lamely look on, Draupadi sees no other recourse than to appeal to Lord Krishna for protection from her assailant.

Krishna responds by granting her divine *darshan* in the Kaurava court, and in typically wily fashion magically provides layers of clothing underneath the one being stripped away by her attacker. The Kaurava brothers furiously escalate their efforts, only to collapse from exhaustion, as they are no match for Krishna's divine powers, which rescue Draupadi from public humiliation. Enraged by the assault, she then summons her husbands and their families to avenge this outrage with the sword. The humiliating episode of the game of dice has a lasting impact in subsequent segments of the *Mahabharat* narrative, leading to a major battle in which the Pandavas triumph over the Kauravas at the conclusion of the epic.

Significant in its narrative intrigue and emotional appeal, "The Game of Dice" is considered one of the most awe-inspiring and troubling segments of *Mahabharat*. It raises many interesting questions about not only narrative conventions of the epic but also crucial concerns about dharmic morality in Indian, or more precisely Hindu, culture. For instance, the question of why the noble Yudhistra, ignoring the well-being of his loved ones, willfully loses everything in the game of dice is never quite addressed in many modern tellings of the epic, including the television *Mahabharat*. Similarly, the issue of how the entire Bharata clan could willingly remain mute witnesses to the humiliation of Draupadi in public view of the court remains disturbing, to say the least. Finally, in Draupadi's burning desire to avenge her humiliation at any cost, the narrative points to an uneasy mix of sympathy for, and accusation against, her role in the ultimate decimation of the Bharata clan.

Given the narrative ambiguities of the "Game of Dice" episode, various tellings of the epic over the centuries have attempted to address the many troubling questions it raises in terms of the cultural morality of their own times. The television *Mahabharat*—like many other modern tellings in literature, films, and comic books—struggles to find a plausible modern rationale for the Pandavas' loss to the Kauravas in the game of dice that leads to the public disrobing of Draupadi. With creative use of the narrative conventions of televisual discourse, the television *Mahabharat* exaggerates Sakuni's cunning manipulations in extreme close-up and represents Duryodhana's arrogant ambitions with melodramatic flourish to effectively delegitimize the Kauravas' victory and generate audience sympathy for the Pandavas. On the other hand, the moral dilemmas raised by Yudhistra's loss in the game of dice are underplayed and are quickly explained away in terms of an uncharacteristic weakness for the game of dice in an otherwise unblemished character of the Pandava prince, who is constantly hailed in the scripted dialogue as "Dharma Raj"—the prince of dharma.

I argue that by representing Yudhistra's willingness to lose everything, including himself and his own family, as an uncharacteristic weakness for gambling, the television series *Mahabharat* seeks to repress archaic traditions of the game of dice by recasting the dharmic epic in terms of modern ideologies of nationalism that value notions of productivity and accumulation of wealth as the nationalist duty and the moral obligation of every citizen. To critique the television *Mahabharat*'s recasting of the traditional meanings of *dharma* in terms of the modern virtues of nationalism is not to suggest that productivity and accumulation of wealth are not desirable traits in a community. To the contrary, to criticize the missionary zeal with which the *Mahabharat* serial recasts modern nationalism as the dharmic duty of every citizen is to bring into question the institutional and ideological power of television in containing and contesting troubling questions about patriarchal authority, cultural morality, personal character, and gender relations in post-colonial India.

These ideological and institutional transformations in Indian television have been associated by Purnima Mankekar with the emergence of a "national family" of upwardly mobile, middle-class consumers who religiously tuned in to Doordarshan's entertainment programming in the 1980s.[35] She argues that national programming on Doordarshan enabled the state-sponsored network to identify with this powerful class of consumers by promoting and participating in the struggle for hegemony over representations of the New Indian Woman—*Nai Bharatiya Nari*. In television serials such as *Udaan* and *Rajani*, the New Indian Woman is represented as a secular, modern, educated, middle-class consumer—but almost always marked as a Hindu. As the idealized representative of the new Indian woman, she is capable of working both inside and outside the home, but she is ultimately devoted to her family and her nation.

Mankekar also interrogates the representations of gender in the "national family" of Doordarshan in two nationalist serials: *Param Veer Chakra* (1990) and *Tamas* (1988). While *Param Veer Chakra* (*PVC*) was a fictionalized account of military heroes who have won India's highest medal of honor, *Tamas* was a dramatic serialization of the tragic tale of the Partition in 1947 and of communal violence between Hindus and Muslims in India and Pakistan. A common theme in these two serials on Doordarshan is the state-sponsored anxiety to recast the violence of nationhood in terms of masculine ideals of patriotic duty and honor, which are constantly mediated by idealized representations of the New Indian Woman in the national family of Indian television. As Mankekar rightly concludes, in the contested terrain of postcolonial nationalism, Doordarshan's idealized representations of women as selfless mothers and loyal wives work effectively to marginalize other female roles that are considered threatening to the national family, and therefore unbecoming of the New Indian Woman.

In her analysis of the gendered representation of the "national family" in *Mahabharat*, Mankekar focuses on the public disrobing of Draupadi by the Kauravas at the end of the "Game of Dice" segment of the epic. Although

Mankekar's research (*Screening Culture, Viewing Politics,* 1999) found that male viewers often identified the final battle scene in *Mahabharat* as the climax of the epic, in-depth interviews with women in Delhi revealed that they almost uniformly fixed their primary attention on the issues generated by the disrobing incident. This was especially true of young, unmarried women, who saw it as embodying their concerns about marriage, dowry, and relations with future in-laws. Men were no less disconcerted by the scene, but for them the issues were somewhat different, involving family pride and indignation over the public assault on a female clan member by an outsider.

The male producers of the television serial perhaps best captured this sensibility when they contended that "the status of women is the most revealing index of how 'enlightened' a particular society is." They saw Draupadi as "the personification of power and energy" in Indian (or more specifically, Hindu) society. Although the producers and female viewers offered different interpretations of significant elements of the serial, a distinctive notion of femininity prevailed among both groups, and it centered around the belief that the power of Indian woman must be contained in order to fulfill the dharmic destiny of the nation, the family, and the community.

During the general elections of 1989, the BJP strategically appropriated the gendered discourse of *Mahabharat* to promote its political ideology of Hindutva as the dharmic duty of every man, woman, and family in the national community. It also drafted some of the leading actors and actresses in the television series as candidates, who then campaigned in costume, winning a number of parliamentary seats. Thus, in rather unmistakable ways, one can see how the BJP's electoral strategy of using the popularity of the *Mahabharat* serial connects with its earlier appropriations of the *Ramayan* serial to further the politics of Hindutva as an alternative to secular nationalism in postcolonial India. Like the public disrobing of Draupadi in full view of the Bharata family, the political marginalization of the Hindutva sentiments in secular nationalism was characterized as a violation of age-old cultural traditions in India, which the BJP has been keen to portray as distinctly Hindu.

To counter the growing opposition to the Congress Party, the Rajiv Gandhi government tried to manipulate Doordarshan's coverage of the various political parties and the major issues and controversies during the election campaigns. Despite the use of state-owned media like Doordarshan to promote Rajiv Gandhi's image, the Congress Party was defeated in the elections, as the government was plagued by allegations about bribes paid to the prime minister himself by a Swedish company, Bofors, in exchange for a contract to supply guns for the Indian military. After the shocking defeat of the Congress Party in the general elections, a fledgling party called the Janata Dal came into power, and its leader, Vishwanath Pratap Singh, once a close associate of Rajiv Gandhi, became the nation's prime minister. V. P. Singh, who had campaigned to rid the government of corruption and bureaucratic mismanagement, also promised to provide autonomy to the state-sponsored media if he was elected to power. Upon assuming office, the Janata Dal government introduced a new bill for the creation of an autonomous broadcasting corporation called Prasar Bharati (in Hindi, *prasar,* "disseminate"; *Bharat,* India).

Disseminating India

The Prasar Bharati Bill was a significant milestone in the longstanding quest for "autonomy" in Indian television. However, the quality of "autonomy" given to Prasar Bharati was rather diluted in comparison to what had been advocated in 1978 under the rubric of "Akash Bharati." The most significant difference was that the Prasar Bharati Bill required the creation of a parliamentary committee to oversee the functioning of the corporation. While critics of this move saw this change as an attempt to curtail the autonomy of Prasar Bharati, the minister for Information and Broadcasting, P. Upendra, assured members of Parliament that safeguards were "built into this Bill with the objective of giving to media the fullest protection from outside interference."[36]

Shortly after the passage of the Prasar Bharati Act in 1990, the Janata Dal Party suddenly fell out of power, before the government could issue the necessary notification to enable the legal creation of an autonomous corporation for broadcasting in India. The party that replaced the Janata Dal government was a minority splinter group led by Prime Minister Chandra Shekar, and it depended heavily on the Congress Party's strength in Parliament to stay in power. Needless to say, the Prasar Bharati Bill was shelved without further action, and broadcasting in India continued to remain under direct state supervision. When the Congress Party withdrew its support for the Chandra Shekar government and forced new elections in 1990, this time its election manifesto contained a promise for the creation of a commission to study anew the question of autonomy for television in India.

In the general elections of 1990, which were clouded by the tragic assassination of Rajiv Gandhi during a campaign visit to Tamil Nadu, the Congress Party was voted back into power. As a shell-shocked nation mourned the death of the Congress Party leader, a new government, led by Prime Minister P. V. Narasimha Rao, was sworn into office. Although the Congress Party government was invested in relaxing several government restrictions on private businesses operating in the television industry in order to boost the sagging Indian economy, it paid little attention to the question of providing autonomy to the electronic media.[37]

In February 1995, on the eve of the Hero Cup International cricket tournament, the government could no longer avoid dealing with the question of autonomy for Doordarshan, after a petition was filed in the Kolkata High Court challenging the authority of Doordarshan's monopoly over television broadcasting in India. The organizers of the Hero Cup, the Cricket Association of Bengal, sold the worldwide rights for telecasting the matches to Transworld Image (TWI) after failing to negotiate a mutually acceptable contract with the state-sponsored network Doordarshan.

Subsequently, the Ministry of Information and Broadcasting asserted that Doordarshan had exclusive rights for broadcasting in India and instructed the government-owned telecommunications provider, Videsh Sanchar Nigam Limited (VSNL), to deny uplinking facilities to TWI. The Cricket Association of Bengal filed a petition before the Kolkata High Court challenging the legality of the government's decision to prohibit the use of the airwaves by private companies. To facilitate the timely conduct of the cricket tournament, the Kolkata High Court issued an interim order allowing Doordarshan to broadcast the Hero Cup within India and permitting TWI to telecast the matches to international viewers.[38]

Although the Supreme Court took up the case after the completion of the Hero Cup tournament, the case, involving the Ministry of Information and Broadcasting, the government of India, and the Cricket Association of Bengal, was extremely significant because of the constitutional issues it raised about the role of the government in regulating the use of the airwaves in India. In its landmark judgment, the Supreme Court held that the airwaves are public property that must be used in ways that ensure the expression of a plurality of views and diversity of opinions in the national community. The Court also ruled that the government of India had a responsibility to use the airwaves to advance the citizens' rights to free speech guaranteed by the Constitution.

In its ruling, the Court explained: "The broadcasting media should be under the control of the public as distinct from Government. This is the command implicit in Article 19 (1) (a). It should be operated by a public statutory corporation or corporations."[39] On the question of broadcasting by private individuals and commercial networks, the Court ruled: "The question whether to permit private broadcasting or not is a matter of policy for the Parliament to decide. If it decides to permit it, its [*sic*] for the Parliament to decide, subject to what conditions and restrictions should it be permitted. Private broadcasting, even if allowed, should not be left to the market forces in the interest of ensuring that a wide variety of voices enjoy access to it."[40]

In defining the airwaves as a public property that is free from both state control and commercial forces, the judges pointed to the "danger flowing from the concentration of the right to broadcast/telecast in the hands of (either) a central agency or of a few private affluent broadcasters."[41] The Supreme Court's historic judgment ordering the government of India to "take immediate steps to establish an autonomous public authority . . . to control and regulate the use of the airwaves" intensified the demands for reform in Indian television.[42]

The Ministry of Information and Broadcasting set up the Sengupta committee to suggest revisions to the Prasar Bharati Act, which had been passed by Parliament in 1990 but shelved by the Congress Party government that came to power in 1991. Although the Congress Party headed the government when the Supreme Court issued its order for the creation of a public broadcasting trust, the general elections of 1996 brought a new political formation, led by the BJP and called the National Democratic Alliance (NDA), into power. Led by Prime Minister Atal Behari Vajpayee, the BJP set out to fashion a workable majority to form the government. However, unable to sustain its majority in Parliament, the NDA government fell out of power after only a short stint of twelve days at the helm of affairs.

When another political formation called the United Front formed the government, the creation of the Prasar Bharati appeared inevitable, particularly since many of the constituent members of the United Front had been parties to the Prasar Bharati Act passed in 1990. On September 15, 1997, the government of India issued the necessary notification to turn the Prasar Bharati Act of 1990 into law. With the creation of the Prasar Bharati Board on November 23, 1997, the task of ensuring autonomy for television in India now rested with the fifteen members of the newly constituted public broadcasting corporation.[43] Now legally enshrined as an autonomous corporation, Prasar Bharati promoted the expansion of Doordarshan by adding more channels to the national network as a strategy to compete with foreign and domestic networks in the realms of entertainment, news, and regional-language programming.

Doordarshan's impressive plans for expansion were supported not only by rising advertising revenues but also by the growing diversity of genres in its programming lineup of daytime soap operas, talk shows, top-ten music rotations, and reruns of nationalistic programs like *Discovery of India,* serials based on traditional folklore like *Chandrakanta,* and religious epics like *Ramayan* and *Mahabharat.*

In a market survey published in 1996, the ad agency Lintas reported that commercial revenue in India had been on the upswing since the introduction of satellite television in the early 1990s. The Lintas report notes that within one year (from 1994–95 to 1995–96) the Indian market grew from 7,620 million rupees ($223.5 million) to 10,200 million rupees ($299 million). For the year 1995–96, Lintas estimated, Doordarshan had gained the lion's share of the commercial revenue, by cornering a whopping $196.5 million (out of a total of $299 million). Zee TV was a distant second, with advertising revenues of $64.5 million, and the Tamil language network, Sun TV, gathered a respectable $13 million.[44] From an all-time high of 5,727 million rupees in commercial revenue in 1997, Doordarshan's share slipped to 4,901 million rupees in 1998 and 3,999 million rupees in 1999, as satellite and cable channels began competing with the state-sponsored network for viewers' attention.

At the same time, as the annual reports (1993–2000) of Doordarshan suggest, from 1991 to 2001, the national network—the indispensable ideological cohort of the nation-state—had grown phenomenally to counter the threat of its transnational and translocal competition in India. In terms of its geographic reach, the network covered 14 percent of the nation in 1982, 61 percent in 1992, and 72.9 percent in 1999. With the launch of the DD-International channel on March 14, 1995, the state-sponsored network extended beyond the political boundaries of the national community and reached viewers across Asia. From ten program production centers in 1982 to twenty in 1992, Doordarshan's production facilities had grown by over 100 percent by 1999, to forty-six centers. Similarly, the number of broadcasting transmitters in the national network increased from 19 in 1982 to 535 in 1992 and 1,041 in 1999. While Doordarshan reached a mere 26 percent of the national population in 1982, it reached 81 percent in 1992 and 87.6 percent in 1999 (see figure 3).

The figures for actual and potential commercial revenues were, however, mere speculations in the rapidly changing Asian markets, where unexpected, intervening variables had ruined some of the best predictions and estimates by media analysts. A case in point is the concept of Direct to Home TV (or DTH), which was expected to grow phenomenally after the expansion of the satellite and cable industry in the mid- to late 1990s. In the first half of the

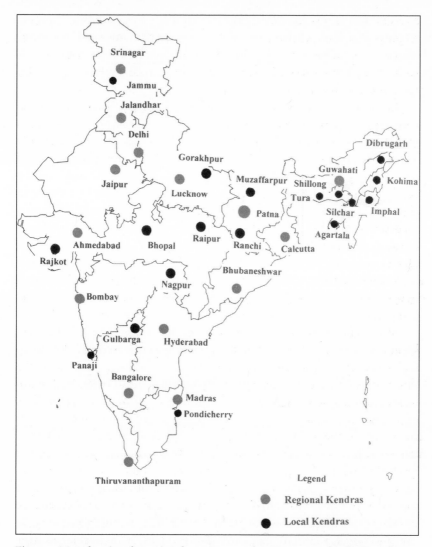

Figure 3. Map showing the regional program production centers for the Doordarshan television network in 1995. Map not to scale. *Doordarshan Annual Report, 1995.*

1990s, DTH was embraced by transnational networks such as STAR TV as a strategy to circumvent their dependence on local cable operators across India, who could drop and add channels from their everyday fare to cater to the programming tastes and linguistic preferences of the viewers in their communities. However, the DTH phenomenon ran into various problems, as the networks recognized that the economics of signal transmission, encryption, distribution, reception, and revenue collection were far too cumbersome in the diverse Asian markets, where, despite half a decade of intensive probing and surveying, media researchers cannot seem to agree even on the number of television households in any one community.[45]

Shortly after the creation of the Prasar Bharati Corporation, the United Front—which had been most receptive to the ideals of broadcasting autonomy—lost power after a short stint of twelve months in government when the Congress Party strategically withdrew its support and called for midterm elections in 1998. After the elections of 1998, the BJP came back to power, but once again, the government lasted for only thirteen months, when one of the regional partners of the NDA, the AIADMK in Tamil Nadu, withdrew its support of the BJP in Parliament. When elections were held in 1999, the NDA alliance gained a comfortable majority to form the government, and it appeared that the third time was a lucky charm for the BJP. However, the future of the Prasar Bharati Act now appeared more uncertain.

In its election manifesto, the NDA had promised to conduct a thorough review of the Prasar Bharati Act after forming the government. What the "review" would result in was not clearly spelled out in the document, but some political analysts predicted that the BJP government was "keen to do away with the half-baked autonomy Prasar Bharati had been provided with."[46] In the run-up to the 1999 elections, the minister of information and broadcasting, Pramod Mahajan, had categorically declared that he was not in favor of granting autonomy to Prasar Bharati as envisioned by the 1990 Act. "I do not have faith in Prasar Bharati. I want Doordarshan under the control of the government. If we come back to power, we will dissolve Prasar Bharati," he proclaimed.[47]

Explaining his aversion to the Prasar Bharati Act, Mahajan had argued that the question of autonomy may have been relevant in 1990 when there was only one national network in Doordarshan, but with the growing number of national, transnational, and translocal satellite and cable channels competing with the state-sponsored network in the 1990s, there was no longer any need for the government to provide "autonomy" to Doordarshan. Instead, Mahajan had expressed interest in focusing his attention, and that of his ministry, toward the development of a Conditional Access System (CAS) to enable the government to better regulate the programming and distribution of satellite and cable television channels in India.[48]

After the 1999 elections, when the BJP-led NDA alliance returned to power, Mahajan was back at the helm of the Ministry of Information and Broadcasting, and he focused his attention on the development of a comprehensive Broadcasting Bill to regulate the activities of the private satellite channels, such as STAR TV, ETV, Sun TV, and Zee TV, and the commercially owned cable companies. The goal of the Broadcasting Bill, introduced in 1997, is to create an independent authority known as the Broadcasting Authority of India to oversee a range of broadcasting services in India.[49]

Given the lack of enthusiasm for Prasar Bharati at the highest levels of government, Prasar Bharati's board members soon realized that their "autonomy" would be severely curtailed by political pressures from the government, coupled with the bureaucratic inertia in the various agencies of the Ministry of Information and Broadcasting. As a frustrated member of the Bharati board, Dr. Rajendra Yadav, once put it, Prasar Bharati is at best a "fractured autonomy."

Is *There an Indian Community of Television?*

If the adoption of the Prasar Bharati Act by the Indian Parliament in 1990 failed to give concrete shape to the notion of autonomy for television, then the creation of the Prasar Bharati Corporation in 1997 did not put an end to the wrangling over legal definitions, political manipulations, and policy confusions in development of public broadcasting. By the end of the 1990s, it became amply clear that the nationalist ambivalence toward granting complete autonomy to broadcasting transcended political ideologies, as three successive governments, led by the centrist Congress Party, the left-wing Janat Dal, and the right-wing BJP, all chose to ignore the core recommendations of the Supreme Court's landmark decision that defined the airwaves as public property. Moreover, as is evident from the policy recommendations of the many government-appointed committees—from the Chanda Committee in the 1960s to the Sengupta Committee in the 1990s—the goals of creating an autonomous corporation for public broadcasting have always been articulated within a framework of a state-sponsored network that marginalizes alternative imaginations of nationalist autonomy in nonstatist terms. In each of these instances, even as the nationalist desire for autonomy in television was being inscribed in the publication of the committee reports, it was simultaneously being erased by the policy decisions taken by successive governments, regardless of the political ideologies or governing philosophies.

Lest we get alarmed by the ambivalence among politicians, policy-makers, and media elites toward the central objective of creating and sustaining the autonomous bodies of Doordarshan and Prasar Bharati in Indian television, we must remember that similar concerns have always been raised about the autonomy of nationalist imaginations in colonial and postcolonial India. In his extensive analyses of print-capitalism in colonial and postcolonial India, Homi Bhabha finds that the ambivalent dissemination of nationalism in popular literary texts like newspapers and novels represent the innermost fears and fantasies of the community about its collective sense of identity in relation to the ever-changing realities of the outside world.[50] For Bhabha, the narrative fears and fantasies of the literary texts, and the ambivalent imaginations of cultural identity and difference they seek to represent, can be only be articulated through a "middle voice" of hybridity. This middle voice, Bhabha argues, is indicated in the double meaning of the term *dissemination,* which, with its hybrid formulation, at once conveys a sense of *fusion with* (the inside/the nation) and *diffusion from* (outside/the world).[51]

At once playful and powerful, Bhabha's appropriation of Jacques Derrida's theory of dissemination (or *differance,* or deconstruction)[52] enables us to articulate the nation *as it is written* in the narrative ambivalence of imagining an "Indian" community of television into existence. However, Bhabha's writings on dissemiNation have extensively focused on print-capitalism and the role of literary narratives of nationalism in the colonial world; they have rarely examined the televisual narratives of Doordarshan and Prasar Bharati in postcolonial India.

In a more recent essay, "Arrivals and Departures," Bhabha finds himself at the "cusp of an electronic dilemma," as he tries to make sense of the new forms of national and transnational communities that are created by the globalization of television.[53] On the one hand, the electronic communities of television are not "national" by definition. On the other hand, the rapid flows of electronic media do not lead us to imagine communities that are completely "detached from national policies of technological innovation, education provision, science policy."[54]

The hybrid character of communities engendered by the electronic flows of television induces Bhabha to wonder "what form of 'media'—in the most general and generic sense of the technology of representation, the genre of mediation—would be appropriate" to the "new rhythms of information and communication" of our time.[55] In what follows, I elaborate on possible answers to the question Bhabha poses by framing my study of the national community of Indian television in relation to what Jacques Derrida has defined as the "techno-tele-media apparatuses" of acceleration and dislocation.[56] What is "accelerated" by the technological apparatuses of television is the ability of users to gain access to a wide range of media and mediated cultures with unimaginable ease. What is dislocated, in this context, "is therefore, a sense of ontology, of the essentiality or inevitability of being-and-belonging by virtue of the nation, a mode of experience and existence that Derrida calls a *national ontopology.*"[57]

Derrida's deconstruction of the radical dislocation of national ontopologies of our time induces us to think of a new mode of nationalist imagination that is now being inscribed and erased—in a flash—by the accelerated flows of televisual technologies and cultures. In this "moment of exposure," as Bhabha puts it, a paradoxical "media temporality" emerges, which shuttles in a double movement to make "*at once* contiguous, and *in that flash*, contingent, the realms of human consciousness and the unconscious."[58]

To address the "erasure within exposure" of the "familiar, domestic, national and homely" in Indian television, I recast Bhabha's literary notion of dissemiNation into a televisual formulation of imagiNation. As a hybrid articulation of the ambivalent dissemination of a nationalist ontopology in Doordarshan and Prasar Bharati, imagiNation thus refers to a "middle vision" of television that is always partially visible in postcolonial India.

As is evident from the recommendations of the Chanda Committee in the 1960s, the Verghese Committee in the 1970s, the Joshi Committee in the 1980s, and the Sengupta Committee in the 1990s, the middle vision of imagiNation has always been partially visible in the policy debates over autonomy in Indian television. For instance, in the early 1960s, when television was emerging as a new medium with distinct potential, in contrast to All India Radio, the Chanda Committee Report proclaimed:

> The question still remains whether the new organisation should follow the patter of AIR and become an attached office of a Ministry or whether it should be made an autonomous, corporation, created by a special statute of Parliament. Having carefully considered the question in all its aspects, we have come to the convulsion that to develop on correct lines television must not be hampered by the limitations of a department; it should have a broader outlook, greater flexibility and freedom of action which the corporate form alone can give.[59]

Although the Chanda Committee Report called for an autonomous corporation for television, when general television services were launched on August 15, 1965, the government of India gave the important task of producing daily, one-hour transmissions to officials at All India Radio in Delhi. In doing so, the government's broadcasting policy not only subsumed the autonomy of television under the institutional structure of All India Radio but also established a central role for the national government in the everyday functioning of the electronic media. Similarly, in the 1970s, as television remained under the centralized authority of All India Radio in Delhi, the Verghese Committee argued for the creation of an autonomous national broadcasting corporation with a highly decentralized structure, and explained its rationale in the following terms.

> The Working Group are [*sic*] of the view that there should not be autonomous regional corporations or even a federal of State Government corporations. Instead, a single National Broadcast Trust is proposed under which a highly decentralised structure is envisaged. There will be a large measure of power delegated to the regional and local level so that the organisation enjoys the advantages of quick decision-making, sensitivity to local problems, familiarity with local customs and taste, and close linkages with various governments and institutions.[60]

The recommendations of the Verghese Committee represented a promising start on paper, but its concept of a single broadcasting corporation was heavily dependent on the ability of governmental institutions to delegate authority to lower levels in the echelons of power. Therefore, the autonomy of the Akash Bharati Corporation that the Verghese Committee proposed was prone to political interference by officials in the governments at the central and the state levels. While policy debates of the 1960s and 1970s centered around institutional control of infrastructure and hardware, the 1980s signaled a new phase in the quest for autonomy in television, as the Joshi Committee recommended a shift in emphasis from hardware to software. Invoking the names of nationalist leaders like Mahatma Gandhi, Jawaharlal Nehru, and Rabindranath Tagore in support of its proposal to rethink the question of autonomy in terms of software or programming content, the Joshi Committee wrote:

> Software planning is of fundamental importance . . . from the point of view of making people aware of the question raised by Tagore, Gandhi and Nehru which has greater relevance to-day [*sic*] than in the past: *Wither India?* Is India condemned to be merely imitative, ultimately reproducing here an inferior version of the Western culture and civilisation? Or can India be creative, building a new pattern combining the best elements of the modern and traditional cultures?[61]

Although the Joshi Committee Report's recommendations emphasized the need for software planning in the creation of an "Indian" personality for television in the 1980s, ironically enough, some of the most creative software planning in Indian television emerged after 1991, when commercial satellite television channels began experimenting with several programming genres such as music television, soap operas, sitcoms, and reality TV shows. As private satellite networks like STAR TV, Sun TV, ETV, and Zee TV began competing with the state-sponsored network, Doordarshan, in the 1990s, policy-makers and politicians alike were forced to confront the rapidly changing environment of television in India. Explaining the need to reevaluate the Prasar Bharati Act, which was passed by Parliament in 1990, the Sengupta Committee argued:

> A complete rethinking of the role, organisation and functions of Prasar Bharati became necessary in a multi-channel scenario, mostly driven by market forces. Prasar Bharati needs the requisite degree of flexibility and financial powers to hold its own. There has been a constant debate concerning the quality and purpose of Indian Broadcasting for quite some time now. Some basic questions will have to be addressed to be able to evolve a vibrant and versatile model of a national broadcasting system, including a reinvigorated Prasar Bharati, in a vastly changed and fast-changing scenario.[62]

The question of whether policy-makers and politicians are able to evolve the "vibrant and versatile" model of pubic broadcasting called for by the Sengupta Committee Report in 1996 remains as yet unanswered. But the "fractured autonomy" that has resulted from the establishment of the Prasar Bharati Corporation in 1997 is yet another instance in the long history of ambivalent policies that shuttle between a nationalist desire to create an autonomous corporation for television and a state-sponsored agenda of maintaining a centralized authority over broadcasting in India. In the process, however, a fleeting vision of autonomy—which I describe as the middle vision of imagiNation—is revealed not only in the policy debates about an "Indian" personality for television but also through the changing structures and hierarchies of broadcasting institutions such as Prasar Bharati and Doordarshan. The hybrid articulation of imagiNation and autonomy is also partially visible in a variety of nationalist programming on Indian television: in the passionate erection and annihilation of historical narratives in docudramas like *Bharat Ek Khoj* that are promoted in the name of national integration; in the artificial insemination and dissemination of the seeds of mythic origins in television serials like *Ramayan* and *Mahabharat* that are celebrated in epic imaginations of the nation; and in the prodigal production and reproduction of the pedagogical instruments of the national state apparatus through performative roles of everyday characters in innumerable sitcoms, soap operas, and dramas.

In each of these instances, what remains most contentious is the question with which I began: "*Is* there an Indian community of television?" Is it an imagined community of cultural production or is it a given product of political-economic institutions? Is it defined through the terrain of the nation-state or is it defied by the transgressions of transnational networks? Is it the residue of archaic traditions or the emergence of a new cosmopolitanism? Is it dominated by the hegemony of nationalist institutions or is it contested by the aspirations of vernacular communities? Perhaps, in each of these questions, both alternatives are at once true and false. Therefore, the question of an Indian community of television, I conclude, can be best addressed in a middle vision of nationalism, for which my preferred term is imagiNation.

2   At Home, In the World: The Viewing
      Practices of Indian Television

IN *PROVINCIALIZING EUROPE*, Dipesh Chakrabarty argues that Benedict Anderson's influential notion of nations as imagined communities is a useful reminder that imagination is a very real and productive phenomenon in everyday life, and therefore should not be understood as something that is false or unreal.[1] Although the central argument of *Imagined Communities* cautions us against reading imagination to mean false, Chakrabarty finds that Anderson takes its meaning to be self-evident. Yet in European thought—which is Anderson's starting point in the history of imagined communities—the meaning of the word "imagination" has a long and complex genealogy. For Chakrabarty, the historic debate over the status of imagination in European thought can be encapsulated in terms of the following question asked of the Spinozan tradition by Coleridge in his *Biographia Literaria* (1815–17): "Was God a subject endowed with a (mental) faculty called 'imagination,' or did God exist simply in the ways of the world without being gathered into anything in the nature of a subject?"[2]

Chakrabarty contends that in the modern history of Western thought—from Coleridge to Anderson—the notion of imagination remains a subject-centered activity of representation through the human act of seeing. The Spinozan tradition of imagination as a nonsubjectivist vision of the divine and the nonhuman, on the other hand, has been relegated to the margins of Western thought. In the Indian context, however, there is a "family of viewing practices" that permeates the mainstream imagination, and that has always been able to reconcile the (Western) split between subjectivist and nonsubjectivist dimensions of imagination. One such family of viewing practices is *darshan*.

A polysemic Sanskrit term that has been borrowed into many north and south Indian languages, *darshan* can be read to mean both (human) seeing and (divine or nonhuman) vision. In articulating the double meaning of imagination as a representation of subjectivist sight and a nonrepresentational, nonsubjectivist vision, Chakrabarty suggests that *darshan* "refers to the exchange of human sight with the divine that supposedly happens inside a temple or in the presence of an image in which the deity has become manifest (*murati*)"[3]

Chakrabarty's insightful discussion of imagination (*darshan*) is restricted to viewing practices of literature in Bengali prose and poetry in colonial discourse. Yet I find it particularly relevant to discuss the multiple meanings of imagination in relation to the family of viewing practices engendered by television in postcolonial discourse. As I have shown in chapter 1, the development of the Indian national network, Doordarshan, has been one of the most imaginative attempts to articulate a family of viewing practices that reconciles the schism between the public and the private, the inside and the outside, the material and the spiritual, the human and the nonhuman in postcolonial nationalism.

However, before extending Chakrabarty's discussion of the polysemic notion of *darshan* to the viewing practices of Doordarshan in Indian television, it may be necessary to distinguish the electronic dimensions of televisual imagination from its meanings in print-mediated literatures. While technologies of print-communication, such as literary prose and poetry, enable a family of viewing practices for the mediated exchange of human sight with divine vision (*darshan*), I argue that the technologies of telecommunication, such as broadcasting and cable television, engender an electronic family of televiewing practices for the immediate exchange of human sight with nonhuman vision (*door-darshan*).

Here, I draw attention to the prefix *door* (meaning tele or distant) in the term *door-darshan,* which, quite literally, refers to electronic dimensions of the immediate exchange of (human) sight with (nonhuman) vision. As a technology of telecommunications, the television set thus transforms from being an idiot-box in the corner to a secular temple of electronic images in which human imaginations and divine visions become manifest (*murati*). In making manifest the dual notion of human seeing and nonhuman vision, television thus engenders an electronic family of viewing practices that transforms the private space of the home into the public stage of the world.

In this chapter, I address the question "Is there *an* Indian community of television?" by deconstructing the viewing practices of *door-darshan* through close textual analysis of print advertisements for television sets in leading national news magazines such as the *Illustrated Weekly of India* and *India Today*. In the first section, I examine some of the earliest advertising strategies used by the Indian electronics industry in the 1970s to promote the viewing practices of television as an electronic exchange of the outside world inside the home. In the second section, I evaluate the changing definitions of the "outside" and the "inside" or the "foreign" and the "domestic" by analyzing advertisements for color television sets made by Indian companies in collaboration with transnational corporations in the electronics industry in the 1980s. In the third section, I focus on the advertising strategies used by the electronics industry during the 1990s as increasing competition among national, transnational, and translocal networks radically transformed collective imaginations of the home and the world in Indian television. I critically evaluate the ways television manufacturers link their respective brand images to the nationalist imaginations and the domestic concerns of their potential buyers.

Since the arrival of transnational corporations such as LG, Samsung, and Sony into the Indian electronics industry in the mid-1990s, I argue, nationalist concerns have been supplanted by globalist ambitions in television advertisements. In the final section, I describe the changing status of the viewing practices of *door-darshan* in terms of a hyperreal construct of the "home theater," which uses new technologies, such as flat screens and surround-sound, to radically transform traditional distinctions such as the inside and the outside, the domestic and the foreign, or the home and the world. I conclude that the question of *an* Indian community of television can best be understood in terms of the viewing practices of *door-darshan,* which seeks to articulate diverse imaginations of nationalism and electronic capitalism in postcolonial India.

Putting the Tele into Vision

Although television was introduced in India in 1959, transmission was restricted to areas in and around the capital city of Delhi for over a decade. In 1965, the Central Electronics Engineering Institute (CEERI) in Pilaini displayed the first indigenously built television set in India. In 1969, the government of India asked CEERI to share its technical know-how with private firms in the electronics industry to promote mass production of television sets. J.K. Electronics in Kanpur was the first Indian company to be issued a license for the production of television sets in 1969. Polestar Electronics and Telerad TV entered the market in 1970, when television services were introduced in the city of Mumbai (then Bombay). In the same year, Weston Electronics and Televista TV also entered the television manufacturing business in Delhi. J.K. Electronics, which had an early monopoly in the television industry, soon emerged as the brand leader, due in large part to the extensive marketing support and the national distribution network of its parent company, the Singhania Group.[4]

A full-page advertisement for J.K. TV appeared in the leading English news magazine of the time, the *Illustrated Weekly of India*, as the company sought not only to establish its brand as "India's largest-selling TV" but also to promote the television set as a cultural commodity that consumers could use at home every day.[5] The headline at the top of the page declares: "The choice of a Queen," followed by the J.K. TV logo imprinted in a box that appears to be a television screen. Beneath the headline and the logo is a photograph of a woman sitting next to a television set with her arm over the top of the cabinet. The ad copy explains, "India's largest selling TV is the first choice of Beauty Queen Farzana Habib, Miss India 1973. You too will be captivated by the elegant appearance and flawless performance of the J.K. Jumbo/Junior" (see figure 4).[6]

As Lynn Spigel has shown in her study of television as a new commodity in the United States of America after World II, representations of women are prominent in the pictorial conventions of advertising that seeks to introduce innovative electronic technologies into the domestic setting of the household.[7] In this context, the relationship between the television set and the beauty queen that is prominently portrayed in the J.K. TV ad suggests that modern women like Farzana Habib have the choice to use television to fight the oppressive norms of patriarchal traditions that deny them access to the outside world by confining them to the domestic space of the home in Indian society.

However, the juxtaposition of a beautiful woman and a television set invites the reader's attention to another relationship: that between the male fascination with the commodification of a new technology in the domestic space and the objectification of a beauty queen's sexuality on the public stage. Yet the manner in which the beauty queen looks directly at the reader indicates that she is in control of her sexuality on the public stage, since the male viewer's gaze is now restricted by television in the domestic space. More

Figure 4. Advertisement for the J.K. TV. *Illustrated Weekly of India*, November 25, 1973, 3.

generally, the blurred boundaries between the domestic space and the public stage are symbolic of the topsy-turvy world of television, which promises to bring home new technologies to "invert the sexist hierarchies at the heart of the separation of the spheres."[8]

Such inversions, perversions, and subversions of the gendered hierarchies between the inside and the outside spaces quickly became a popular theme in advertisements for television sets, as other manufactures like Weston and Televista started aggressively competing for their share in the limited markets of the Indian electronics industry.[9] In 1972, Televista TV launched a national advertising campaign with a full-page color ad.[10] This early advertisement for the black-and-white television sets made by Televista is relatively simple in composition yet stunning in color and overall effect (figure 5).

Following the golden-thirds rule of art composition, the ad can be clearly distinguished into three equal visual zones. The top third of the ad depicts a tree-lined landscape with reddish-golden skies that nicely merge into the background color for the bottom two-thirds of the page. In the center of the top third of the page is the silhouette of a crane (bird) perched against the setting sun. In the center of the page (or the second third) is the image of a Televista television set. On the screen of the television set is the silhouette of a crane against the setting sun—an exact replica of the image that forms the backdrop for the top third of the page. Between the two images—or the mirrorings of one image—is the headline in large text: "It's the inside that is responsible for what happens on the outside." At the bottom third of the page, slightly to the right, the brand is identified in text as "Televista The Big Name in TVs," followed by a listing of available models and the company address. To its left is a textual description of the product that continues the play on the inside/outside motif, as follows.

> Take a look inside a Televista tv.
> You'll see that Televista incorporates all the latest advances made in tv technology.
> Use of integrated circuitry. (ICs)
> The new way to better tv performance.
> New valves. New everything. Now look outside. The pictures come sharp and clear. Motion-picture reproduction. The sound. Superb hi-fi.
> See the finely grained woodwork.
> No laminated imitations. Pure teak. But most important is the after-sales service. At your doorstep. Fast!
> Visit the nearest Televista showroom. And see everything for yourself.[11]

Figure 5. Televista advertisement. *Illustrated Weekly of India*, July 9, 1972, 28.

What is most striking about this earliest of advertisements for television in India is the emphasis on mirroring, which is a characteristic feature of realism in modern societies. Briefly, *realism* refers to a subject-centered imagination of reality wherein the objective world outside is privileged as being authentic and real. Images or representations of that outside reality, on the other hand, are considered *mimesis*, that is to say, inauthentic imitations or mimicry of the real. The aim of representational forms like art—or, for that matter, advertising and television—in modern realism is, thus, to project an idealized mirror image of the reality outside.

In Televista's attempts to promote the viewing practices of television in India, we find that the ad image emphasizes realistic conventions—use of mirroring and inside/outside distinctions—yet it betrays the ambiguity and the arbitrariness of the distinction between the subjectivist representation and the objective world that is inherent in the discourse of modern realism. For instance, in the Televista ad just described, the ad imagery seeks to establish the ability of the television set to electronically reproduce a mirror image of the natural world by using the inside/outside motif of realism. Never mind if the early broadcasts in India were only in black and white; the print advertisement seeks to authenticate the televisual image by representing its synthetic reality in true-to-life color.

What follows the mirroring of imagery in the Televista ad is the headline, which is a further play on the inside/outside motif: "It is the inside that's responsible for what happens on the outside." Interestingly, to emphasize the television's (mirror) imagery, the headline attempts to invert the conventional logic of realism: the image inside television does not merely mirror the reality outside but also is responsible for sustaining the world outside. Finally comes an elaborate statement that seeks to legitimize the inversion of realism, and the tool used to do this is, of course, the new technology of television: "Televista incorporates all the latest advances made in tv technology. The use of integrated circuitry. . . . New valves. New everything . . ."[12]

Suddenly, the ad confronts yet another fundamental contradiction of representation in modern realism: If everything is being created anew through the use of technology, how can television claim to be realistic? Thus, the description immediately relapses into conventional realism: "See the finely grained woodwork. No laminated imitations. Pure teak." Finally, as the description approaches the end of its contradictory claims to realism, the ad is forced to confront the incredulity of its own narrative before the disbelieving reader: "Visit the nearest Televista showroom. And see everything for yourself."[13] The inversion of realist conventions affected in the headline and the ad imagery is quickly reinverted at the end of the narrative in the uncomfortable recognition of the reader's doubts.

As if in recognition of the reader's ability to discern the impossible claims of modern realism, Televista launched another ad that abandoned the quest to authentically represent the outside world inside the television set and instead celebrated the synthetic reality of electronic media by acknowledging the affinities between the televisual screen and the cinematic screen.[14] Remarkably similar in its composition to the first ad with the mirror image of the crane, this ad also can be clearly distinguished into three equal visual zones.

At the top of the page is the headline, which proclaims: "Televista For that Cinema-Screen Reality." The background of the headline merges into the reddish-golden skies, which frame the silhouette of a couple (presumably the hero and heroine of a film) against the setting sun. In the center of the page (or the second third) is the image of a Televista television set. On the screen of the television set is an exact replica of the silhouette of the couple standing against the setting sun. At the bottom third of the page, slightly to the right, the brand is identified in text as "Televista The Big Name in TVs," followed by the company's address. To its left is the following description of the product.

> Televista was designed to be at par with any other set in the world. So you can be sure that only the best goes into it. Improved longer lasting values. Hybrid and integrated circuitry (IC) . . . No wonder, that cinema-screen reality. In addition, Televista offers you a host of other features which few others give. Like smart styling. Superfine tuning for sharp, clear pictures. Hi-fi sound reproduction. Five advanced controls for easy tuning. Width and height controls that ensure steady pictures. A solid one year guarantee. And of course, two great sets to choose from. Playmate and Elite. Both their screens measure 47 cms. diagonally. Which means that you can sit nearer without tiring your eyes and also see better than ever before.
>
> As for after-sales service—which you may never need—it's unbeatable. Just make a phone call. And our service engineer will be at your doorstep.[15]

What is clear from the new emphasis on the televisual imagination of cinematic reality in Televista's advertising strategy is a recognition that moviegoing is the most popular form of entertainment in India. It also demonstrates the eagerness of the makers of Televista TV to exploit the popularity of film-based programming—such as *Chayageet* (and later *Chi-*

trahaar), which featured song-and-dance sequences from Hindi cinema; *Phool Khile Hain Gulshan Gulshan,* which provided interviews and insights into the private lives of Bollywood heroes and heroines; and late-evening feature films—to attract the attention of consumers to its brand of television sets. However, the Televista ad also uses the popular appeal of Indian cinema to promote the novelty of the television set as a desirable commodity for viewers, who could now watch their favorite stars in the comfort of their own homes instead of jostling with the long lines of impatient crowds at the movie theaters.

In *Television: Technology and Cultural Form,* Raymond Williams introduces the concept of "mobile privatization" to make sense of what he describes as "an unprecedented condition" engendered by the rapid rise of television as a popular cultural commodity in the second half of the twentieth century.[16] With the arrival of the television set in the living room, consumers of mass media are introduced to a new family of viewing practices that enables them to imaginatively connect to an increasingly mobile world around them, even as they remain seated in the restricted privacy of their homes.

Recognizing the radical innovation of mobile privatization introduced by television, the Televista ads highlight the mobility and privacy engendered by the viewing practices of *door-darshan* and promise to deliver the "cinema-screen reality" on a squarish screen measuring a measly 47 centimeters diagonally, in comparison to the panoramic views offered in the movie theaters. Since the Televista TV sets have "improved longer-lasting valves" and "hybrid integrated circuitry," the ad confidently proclaims, "No wonder, that cinema-screen reality." The result of these technological innovations is "that you can sit nearer without tiring your eyes and also see better than ever before."[17]

In a series of advertisements that followed in 1973, Televista TV furthered the play on the inside/outside motif and the cinematic/televisual relationship, this time, however, suggesting that the television set can create a new reality at home that is superior to anything else one may encounter in the outside world or on the cinematic screen. "If you have seen the flashing finger-work of Ravi Shankar, would you settle for second-best?" reads the headline for one Televista TV ad that features a television set with an extreme close-up of the sitar player's hands.[18] The sitar itself protrudes out of the television set toward the headline above, as if to suggest that televisual reality can recreate the three-dimensional ambience of the concert hall in the viewer's living room—that too in extreme close-up.

Another ad in the series features a close-up shot of the noted dancer

Yamini Krishnamurti on the television screen, with her hands extending into the space in front of the television set. The headline asks: "If you've witnessed the entrancing artistry of Yamini Krishnamurti . . . Would you settle for second-best?"[19] The headline for a third ad in this series poses the following question: "If you have seen the genius of Dilip Kumar's character-portrayal . . . Would you settle for second-best?"[20] Prominently occupying the bottom half of the page is a close-up of the Bollywood superstar Dilip Kumar holding a woman's necklace in his right hand, which extends beyond the television screen as if in an offering to the viewer at home.

Through a strikingly simple yet powerful theme of the electronic reproduction of the outside world inside the television set, the three ads promote a collective imagination of televisual reality that takes the viewer at home into public venues such as the concert hall, the dance theater, and the cinema theater to provide a close-up view in a way that is even better than actually being there. Spigel describes television's promise to alter the relationship between the stage space of the theater and the domestic space of the home in the following way: "It wasn't simply the image of the stage on the television screen that gave the illusion of being at the live theater; rather it was the alternation between the stage space and the domestic space that produced a sense of 'being there.' Through this alternation, viewers experienced a kind of layered realism in which the stage appeared to contain the domestic space, and thus, the stage seemed spatially closer—or more real—than the domestic space."[21]

In this sense, the headline "Would you settle for second-best?"—which is a blatant promotion of Televista's market edge over its competition—could also be read as a clever attempt to elevate the "close-up" views provided by television as better viewing practices than those available at other entertainment venues like cinemas, theaters, and concert halls. It is important to remember that television in India was a novelty in the early 1970s, and television services were only available in and around three or four cities across the country. A survey published in December 1974 to evaluate popular reaction to the new medium of television in Mumbai estimated that the number of television sets in India was no more than 140,400 (85,000 sets in Delhi, 55,000 sets in Mumbai, 150 sets each in Poona and Srinagar, and 100 sets in Amritsar).[22] In this limited market for television sets, major manufacturers like Televista used aggressive advertising campaigns to promote their brand popularity over other nationally recognizable names like Weston and J.K.

In 1975, Televista found itself slipping into the unenviable position of being the "second best," when Weston TV took over the mantle of market leader, in a year when sales had fallen drastically for all television manufacturers. In 1976, the major and minor players in the electronics industry banded together to form the Indian Television Manufacturers Association (ITMA) and lobbied the government for lower taxation and excise duty on television sets. When the government of India agreed to provide incentives for companies making low-cost television sets, there was intense competition among manufacturers to cut prices.

Ironically enough, many small-scale businesses manufacturing low-cost television sets collapsed in this competitive environment, as the major players in the electronics industry were able to slash prices on their economy models and make up the deficit by raising the prices of their deluxe models. However, not all the major players in the television manufacturing business survived the shakeup in the electronics industry. When J.K. Electronics was forced to quit the business in 1977, Weston and Televista further consolidated their positions as the first- and second-placed companies in the electronics industry.

It is important to recognize that even the largest companies in the Indian television industry in 1977, such as Weston and Televista, were technically "small-scale" businesses, since their production capacity was no more than 10,000 to 20,000 sets per year. However, relative to other small-scale businesses, which produced around 1,000 to 5,000 television sets per year, Weston and Televista were certainly large-scale enterprises. Moreover, the larger companies, which manufactured a variety of electronic goods such as cassette tape recorders, transistor radios, and car stereos, could sell their television sets in the economy range for almost no profit by spreading out their costs across the entire product line.[23]

In an ad that appeared on the inside page of the front cover of *India Today,* Weston—the self-appointed "electronics people" of India—unleashed a range of products, including television sets, transistor radios, cassette tape recorders, car stereos, and electronic calculators. "The Weston Range"—as the headline put it—was prominently arrayed in the middle of the page with the Himalayan mountain range in the backdrop, suggesting that the company had attained stratospheric heights as the brand leader in the television manufacturing business in India.[24]

Televista, now the second-best-selling brand, also unveiled its range of products in a full-page advertisement that declared: "QUALITY is the most important component we pack into all our products. . . . That's what makes them different."[25] Other television manufacturers were not to be left behind;

for example, Crown TV promoted its collection of electronics with similar claims of "unmatched quality."[26] Crown TV's ad copy explained:

> The Crown Collection. Whether it is our televisions, our cassette recorders or our two-in-ones. There is one thing they all have in common. Unmatched Quality.
>
> That is because intensive research and advanced technology has gone into the construction of every set. And each set is critically inspected at every stage of development. To give you peak performance and trouble-free service.[27]

A common thread in the aforementioned advertisements for Weston, Televista, and Crown is the growing concern over the rapid transformation of the electronics industry in India during the 1970s, as the brand leaders fought gamely to keep abreast of technological innovations and stay one step ahead of their competition. Although the headline and the text in each of these ads seek to emphasize the superior quality of the electronics that sets apart one brand from the rest, it is important to note that none of the major companies in India (including Weston, Televista, and Crown) actually produced any of the components used in the television sets.

Since foreign companies were banned from participating in the television manufacturing business in India, components like transistors and integrated circuits used in the TV sets were produced by various small-scale electronic companies in the private sector. At the same time, many state governments entered the television manufacturing business in an attempt to promote the growth of their electronics industries. For instance, the Electronics Corporation of India (ECIL), based in the state of Andhra Pradesh, manufactured television sets under the brand name EC TV, and the Tamil Nadu Industrial Development Corporation, in collaboration with Dynavision in Madras, produced Dyanora TV sets.

The competition between the government-sponsored undertakings in the public sector and the leading television manufacturers in the private sector was quite fierce, as is demonstrated by the following headline in an advertisement for Dyanora TV: "Why can no other manufacturer risk saying all this?"[28] The ad copy under the headline offers the reader a list of "12 vital promises," which, the makers of Dyanora inform the reader, "cover, between them, everything you need to look for in a TV."[29] At the end of the ad in the right corner of the page is a small image of a television set, above which is the slogan "Take the advice of your eyes. And ears. Dyanora TV. Solid State."[30]

The solid-state technology promised in the Dyanora TV ad was one of the more advanced technologies used by manufacturers of television sets in the late 1970s. Television sets introduced in the early to mid-1970s were based on a hybrid design that combined vacuum tubes, valves, and transistors. Since vacuum tubes and valves require large amounts of electricity to operate, they generate high quantities of heat, which sometimes leads to the failure of other electronic components in the television sets.

The solid-state technology was a significant advance over the hybrids, since it consumed less electricity and produced very little heat. Moreover, television sets using solid-state technology also produced a clearer picture and better sound, as is alluded to in the Dyanora TV slogan "Take the advice of your eyes. And ears." After the introduction of solid-state technology by public-sector undertakings like ECIL and Dynavision, the major companies in the private sector, like Weston, Televista, and Crown, followed suit.

Very soon, however, the television manufacturers were swept away by the next generation of electronics, including integrated circuits (IC), medium-scale integrated circuits (MSI), and large-scale integrated circuits (LSI). As a result of the high demand for the latest technologies in television sets, the electronics industry in India began to experience dramatic shortfalls in the raw materials necessary for manufacturing components. Since many of the raw materials—such as high-quality glass for picture tubes and television screens and transistors for circuitry—were not available in the domestic markets, the government of India was forced to remove some of the import restrictions on electronic components to meet the increased demand for new television sets. At the same time, however, the government also imposed higher rates of excise duty on foreign imports in the electronics industry as a way to regulate the foreign exchange reserves in the national economy.[31]

In 1979, when the government of India raised the excise duty on imported electronics, there was a sharp increase in the price of even the most basic economy model called the "Janata" set, which was promoted as "the television set for the masses." When the Janata television sets were subject to 15 percent excise duty, it represented a 10 percent increase from the previous year. At the same time, excise duty on the "deluxe" television sets was raised from 21 percent to 30 percent, putting them beyond the reach of most middle-class and low-income families in India. To decrease the electronics industry's dependence on foreign imports and increase the availability of low-cost television sets in the market, the government commissioned Bharat Electronics Limited (BEL) to indigenously produce picture tubes—the component accounting for about 40 percent of the total production costs in a monochrome television set.[32]

However, the picture tubes made by BEL cost three times more than imported ones, and thus the Indian government's policy provided very little incentive for television manufacturers to use the indigenously produced component. Therefore, many television manufacturers started using unauthorized foreign circuitry in their sets, and the electronics industry slowly shifted its focus from indigenous production to increased reliance on imported technologies. Due to the illicit nature of their unauthorized use of imported electronics, most manufacturers were loath to advertise the virtues of foreign components in television sets.

On the other hand, manufacturers who were still using homemade electronics in their television sets were only too keen to point out that difference, as evidenced by an advertisement for Bharat TV with the following headline: "When all the novelty wears off, you're left with the TV for India."[33] The ad copy underneath the headline cautions consumers not to put their faith in the gimmicks used by competing manufactures who promise that their television sets are "ultra sophisticated," "integrated," and "revolutionary." Why? "Simply because Indian conditions are unsuitable for them." The television sets built at Bharat Electronics, the ad informs the readers, are "Tried, tested and triumphant in India."[34]

One of the most interesting features in this ad is, of course, the creative way in which the makers of Bharat TV try to delegitimize both the authorized and the unauthorized uses of imported circuitry without a mention of the "foreign" components in the television sets of competing manufacturers. What enables this strategy to work in this particular ad is the use of nationalism as a unique selling point of Bharat TV (*Bharat,* "India").

However, nationalism is invoked here not through an imagined sense of patriotism and cultural pride but through a pragmatic appeal to the purses and pocketbooks of consumers by acknowledging the uniquely "Indian" problems of televiewing practices in the hi-tech world of electronics. "Take any common complaint of a TV owner in India; you can be sure that Bharat TV has an answer for it," the ad assures the consumer who is concerned about the problems of breakdown and the high costs of repairs.

By an unsaid corollary, the makers of Bharat TV attempt to remind the reader that the foreign manufacturers of electronic components—however advanced and sophisticated—do not quite understand the market conditions in India and thus are unable to address the domestic concerns of the Indian consumer. In spite of the creative advertising strategies used to promote Bharat TV as the "television for India," an indigenously produced television set was still prohibitively expensive in large part because of the high costs of raw materials in the electronics industry and the heavy taxation and excise duty on imported components by the government of India. In 1979, Bharat Electronics Limited (BEL) announced its plans for the production of an indigenously produced Janata set with a retail price tag of 1,400 rupees and offered to share its know-how with small-scale manufacturers of low-cost television sets.

Despite the technological advances made by the public-sector companies like BEL, many of the home-grown electronics in India were, as Subhrajit Guhathakurta puts it, "cumbersome and unpredictable."[35] Moreover, he argues, the high failure rates of the television sets made it very difficult for many manufacturers to cope with the servicing demands of their customers, and several small-scale enterprises went out of business by the end of the 1970s. Between 1973 and 1978, while the number of television sets in India grew from 70,000 to 410,000 the number of television manufacturers increased from twelve to seventy-two. Moreover, the manufacturing business was completely dominated by the twenty largest firms, which accounted for 99.5 percent of the sets sold in 1973 (which only dropped to 95.1 percent in 1977).[36]

The four largest firms had a market share of 61.18 percent in 1973, which dropped in subsequent years but remained substantially high, at 48.07 percent, in 1978. Among the reasons for the uneven transformations in the television manufacturing business during the decade between 1970 and 1980 was a state-sponsored ambivalence—what Kumar describes as a policy of "to be or not to be."[37] Since there was no consensus on the question of whether television sets were "common media for mass communication" or "a necessary luxury to denote progress" or "a technological must" for the nation to build its audio-visual infrastructure, the result was a "tussel [sic] between those who are all for television . . . and those who are against" it.[38]

Bringing Home the World

In May 1981, the government of India set up a Working Group to recommend a comprehensive strategy for the introduction of color transmission accompanied by the indigenous production of color television sets in time for the Asian Games in 1982. The Working Group recommended that CEERI take the lead in the production of color television sets and the necessary components and share its technical know-how with small-scale manufacturers in the electronics industry. In January 1982, CEERI produced its first prototype for a fifty-one-inch color television set, and six months later, in July 1982, a demonstration of the CEERI models was organized for the prime minister.[39]

However, given the looming deadline of the Asian Games, the government of India shelved its plans for indigenous production in the small-scale sector and allowed the electronics industry to import color television kits for quick assembly and rapid distribution across the nation. Manufacturers of indigenous television sets were strongly opposed to this change in policy, since it was assumed that it would benefit major players in the electronics industry who had the necessary infrastructure to rapidly assemble, distribute, and market the new color television sets. Not surprisingly, Weston, the undisputed brand leader in the television industry, announced "India's first Color TV."[40] Pointing out that Weston was "the authorised service centre for Hitachi, Japan," the ad copy proclaimed:

> Weston, the pioneer of entertainment electronics in India, brings you a world-class color TV. Assembled by highly qualified engineers trained at HITACHI, Japan. And tested on sophisticated instruments from HITACHI. Backed by 52 full-fledged service centres to give you prompt after-sale service, wherever you are.[41]

What is remarkable in this ad is the manner in which Weston Electronics seeks to establish its status as the market leader in the Indian television industry (and "No. 1 in color too") by playing up its association with Hitachi of Japan. Other manufacturers who could only envy Weston's ability to sell its "No. 1" standing resorted to an advertising strategy that played on the consumer's desire for an economical option for watching the Asian Games on a color television set, as is apparent in the following text in an advertisement for Texla TV.

> The incredibly low priced color TV with unmatched super features make [*sic*] Texla Techno colour an all time bargain. Add to it the Split Second Service of Texla, with its maintenance and service guarantee, watching the Asian Games is going to be a funfilled and exciting experience.[42]

After the successful completion of the Asian Games in 1982, the government of India decided to review its policy on foreign kits and other imported components and affirmed its support for the indigenous production of color television sets to ensure a level playing field between the major players and the small-scale manufacturers in the electronics industry. However, in a major reversal of national policy in 1985, the government allowed India manufacturers to join forces with foreign companies by liberalizing the import of color picture tubes and other electronic components for television sets. The result of the government's liberalization policies was a spate of collaborations between the major players in the Indian electronics industry and transnational corporations such as Fisher, Hitachi, JVC, Mitsubishi, Phillips, Sony, and Toshiba, to name a few.

As each collaborative venture promoted the unique features of its "foreign" connections, advertisements for color television sets in India now appeared to provide a veritable list of "who is who" in the transnational electronics industry. An advertisement for Binatone Electronics announced that its fifty-one-centimeter Colorama Deluxe model was "Introducing the standards by which Indian Colour TVs will be measured for years to come." Counseling the reader, the ad continued: "So before you buy your colour TV. See Binatone. Then Decide." Emphasizing Binatone's transnational credentials in England, Germany, Japan, Korea, Taiwan, and Hong Kong, the ad copy proudly declared: "Now in India."[43]

Unlike Binatone, which aggressively promoted its transnationality, Beltek TV took a more restrained approach to advertising its Euro-Color TV "crafted with the help of 270 computerised quality control tests, using state-of-the-art technology from ITT of West Germany." Relatively simple in its composition, the ad prominently features an image of a television set with the silhouette of a swan against the setting sun on the screen. Above the television set is the headline "Beltek," and below the television set the text continues: "And the other [brands of TV sets]. The difference is color."[44]

Taking this theme of the superiority of imported electronics further, in another Beltek TV ad, the headline above and below the television set (this one with an abstract image of a color burst on the screen) reads: "Beltek. And the other. The difference is technology."[45] During this period, two of the most exaggerated elements in television advertisements were, to use the Beltek slogan, "A million colors on screen" and "state-of-the-art technology," as the leading manufacturers in the electronics industry exploited the novelty of color to promise an enhanced viewing experience over the black-and-white sets that consumers were familiar with.

In one advertisement for Solidaire TV, the headline boasts: "If the eyelashes of the news-reader are clearly visible . . . it must be Solidaire" (*India Today,* December 31, 1985, 191). The ad features the Doordarshan news-reader Gitanjali Iyer on a television screen, around which the headline is displayed in two parts at the top and the middle of the page, respectively (see figure 6). At the bottom of the page is the slogan "Solidaire—that seldom fails!" Between the slogan and the image of the news-reader are two simple-looking television sets and an oversized remote control, and to their left is the following text.

Figure 6. Advertisement for a Solidaire color television. *India Today,* December 31, 1985, 190.

Yes, a claim Solidaire alone can claim. With confidence. Since it has won acclaim from its proud owners. Leading to make more and more people buy
Solidaire. For its picture clarity, perfect colour, and the right contrast. And to
judge the picture quality, you must do it when the news is being telecast.
Because it's live and the news-reader faces the TV camera under proper lighting. It means there is no chance of bad focussing. Now watch the same on
different TVs and compare it with Solidaire. On a Solidaire TV you can see
every minute detail like the eyelashes of the news-reader. And all in the right,
natural tones.

Such picture sharpness in Solidaire is the result of using error-free components from ITT & Preh, West Germany. For instance, components like the
LOT, Electronic Tuner. Colour Ics, SMPS etc used in Solidaire are among the
most advanced designs in the world.[46]

The headline of the ad is indeed unique in terms of the exaggerated claims
it makes about Solidaire TV's ability to make visible the minutest details of
a news-reader's face. However, the most interesting element of Solidaire's
claims is embedded in the ad copy, which provides a step-by-step manual
for the viewer "to judge the picture quality" by watching television "when
the news is being telecast . . . live and the news-reader faces the TV camera
under proper lighting."[47] Recognizing that color television was a relatively
new phenomenon in India, many manufacturers were keen to highlight their
collaboration with foreign companies to promise consumers "the most advanced designs in the world."[48]

In the Indian electronics industry of the 1980s, downplaying one's foreign
connections was clearly a thing of the past, as television manufacturers vied
with each other to provide the viewers at home with newer imaginations of
the world outside—"and all in the right, natural tones," as the Solidaire TV
ad proclaims. In their efforts to produce state-of-the-art television sets with
"the most advanced designs in the world," the major players in the electronics industry were also encouraged by the Indian government's new policies
toward unrestricted technical collaborations between domestic and foreign
companies. In this context, Subhrajit Guhathakurta summarizes the developments in the Indian electronics industry during the years between 1981 and
1987 as follows.

(1) An overhaul of the duty structure and relaxation of capacity limits. Most
restrictions on entry were removed and many large private-sector enterprises were allowed to compete with small-scale and public-sector concerns;
(2) the introduction of color TV transmission and provision of licenses for
assembling imported color TV kits; (3) deregulation of subscriber-end tele-

communication equipment for the private sector (ending the public-sector monopoly in this industry); (4) preference for fiscal controls (duties, taxes) as opposed to physical controls (total ban on importing some products); (5) import of technology was allowed subject to a "phased manufacturing program" for quick indigenization; and (6) most electronic components for which substantial indigenous capacity was not developed were allowed to be imported freely. Besides these measures, specific policies were formulated for some products that were expected to grow significantly in the future (e.g. color TVs, computers, and computer software). In all cases, the policy package included substantial deregulation and delicensing.[49]

Although the liberalization policies of the 1980s summarized by Guhathakurta in the foregoing passage were a significant departure from the protectionist policies of the previous decade, the deregulations and delicensing procedures initiated by the government were not "a complete turnaround, opening the borders for free trade."[50] Rather, it was a "selective and cautious" process for determining what foreign imports were vital for the development of the Indian electronics industry. For instance, Guhathakurta finds that "no imports of consumer electronic products were allowed except as personal baggage with a 240 percent duty," while "for those components that were not available indigenously or were available in limited quantity, imports were automatically approved," with import duties ranging between 75 percent and 150 percent.[51]

Although the imposition of severe import duties was aimed at protecting small-scale manufacturers and public-sector companies, the government's liberalization policies "allowed all Indian companies, regardless of size (including those with foreign equity participation of 40 percent or less), to operate in any field of electronics," including the television manufacturing business. Moreover, any company that had more than 40 percent foreign equity was allowed to "set up manufacturing facilities for electronic components and sophisticated 'high tech' instruments."[52]

The political rationale for the government's "selective and cautious" embrace of deregulatory policies was driven by an economic calculation that the rising demand for television sets in India would lead to an increase in the indigenous production of electronic components, and subsequently decrease the domestic manufacturers' reliance on foreign imports. Moreover, as discussed in chapter 1, during the 1980s, as the state-sponsored network rapidly spread into cities, towns, and villages across the country, Doordarshan's national programming—beginning with primetime serials such as *Hum Log* in 1984 and *Buniyaad* in 1985 and extending to religious epics like *Ramayan* in

1987 and *Mahabharat* in 1989—captured the collective imagination of Indian viewers. No longer a luxury item for keeping up with the Janardhans in the neighborhood, the television set had become a necessary commodity for staying informed and being entertained.

When Doordarshan announced its plans for the extensive and live coverage of the Reliance Cricket Cup in 1987, many manufacturers resorted to creative advertising strategies to promote the technological innovations of their models over those produced by their competitors. Onida TV, in collaboration with JVC, launched a national advertising campaign to unveil its new vertical television set, which was a major innovation in Indian television. The innovative ad campaign—with the devil as a spokesperson for the company's slogan, "Neighbour's envy. Owner's pride"—created a national buzz for Onida in the overcrowded color television markets in India (see figure 7).[53]

In a short span of two years, the total volume of black-and-white television sets sold in India increased from 2,150,000 in 1985–86 to 4,400,000 in 1987–88. At the same time, sales of color television sets grew from 900,000 in 1985–86 to 1,300,000 in 1987–89.[54] During this period, some of the fastest growing companies in the electronics industry were small-scale and medium-scale establishments run by a heterogeneous group of entrepreneurs with engineering, technical, or managerial skills. In 1984, the Electronics Trade and Technology Development (ET&T) Department of the government of India launched an innovative scheme to provide technical, material, and brand-name support for small-scale manufacturers in the electronics industry. The ET&T plan enabled small-scale establishments "to avail themselves of volume discounts in price and superior quality materials which, previously, only the large-scale manufacturers could command."[55]

One of the medium-scale companies was Calcom, which grew a phenomenal rate of about 85 percent annually between 1981 and 1989 by serving as a supplier of black-and-white television sets to many of the nationally marketed brand names.[56] At the same time, the nationally recognizable players further consolidated their brand names through diversification of their product range in consumer electronics in order to attain an economy of scale that would spread out the costs, minimize risk, and maximize profit potential in a highly uncertain market. This strategy is clearly evident in an advertisement for BPL announcing its product range with the headline "India, sings, watches, talks, diagnoses, copies and computes with BPL."[57] While the ad imagery displays a variety of electronic products from BPL, the ad copy goes further:

Figure 7. Advertisement for the Onida vertical television. *India Today*, January 31, 1990, 66.

Entertaining millions with its TVs, VCRs and Two-in-Ones. Saving lives with
Electrocardiographs and Intensive Coronary Care Units. Running offices
efficiently with Electronic PABXs, Push Button Telephones, Communication
Networks, Plain Paper Copiers, and Computers . . . With a promise of more
to come: Cordless Telephones, Rice Cookers, Microwave Ovens, Dishwash-
ers, Vacuum Cleaners, Washing Machines, Refrigerators.[58]

To establish BPL's reputation as the brand leader in the Indian electronics industry, the ad provides the reader with substantial details about the
company: "Rs. 450 crores [*crore* means "ten million"] turnover, 40 products,
1,000,000 square feet of built-up factory space, over 4000 employees, 4 R&D
Centers, 26 Sales and Service Centers, 2500 dealers, exclusive technical arrangements with Sanyo of Japan."[59] As highlighted in the ad copy, BPL set
up several factories to produce all the electronic components (with the ex-
ception of picture tubes, which had to be imported due to quality concerns)
in order to attain the vertical integration necessary for the brand leader to
stay ahead of its closest competitor in the television manufacturing business,
which was beginning to slow down by the end of the 1980s. The volume of
color television sets sold decreased to 1,150,000 in 1988–89 from a high of
1,300,000 in 1987–88. At the same time, the sale of black-and-white television sets in India declined to 3,500,000 in 1988–89 from the higher figure
of 4,400,000 in 1987–88.[60]

"Taking the Tele out of Television"

The arrival of foreign satellite networks, led by STAR TV in May 1991, provided a vast new potential for the production of new television sets in the
Indian electronics industry. Many of the black-and-white and color television
sets sold in the 1980s had the capacity to receive only eight to ten channels,
and they did not have the S-Band tuner to receive the satellite networks now
available through the local cable operator. Moreover, many of the low-end
budget models manufactured in the 1980s did not include a remote control,
since most viewers did not find it a necessity when Doordarshan was the
only network they could tune into at home. However, as cable television
spread across major cities and towns across India, many viewers found it
necessary to buy a new model with a remote control that would enable them
to flip through the various satellite channels without getting up every time
to turn the dial on the television set.

The leading players in the Indian electronics industry, such as BPL, Onida, and Videocon, developed a range of cable-ready television sets both for the economically concerned consumer and the luxury-oriented buyer. In 1991–92, Videocon emerged as the brand leader, capturing 22 percent of the Indian television market, followed closely by BPL, with 21 percent, and Onida, with 20 percent. This trend continued until 1994, when Onida's share slipped to 17.5 percent, due in large part to the aggressive price-slashing strategies of Videocon and BPL to increase the sale of their low-end models, coupled with innovative advertising strategies to launch new models at the high end of their price range.[61]

In 1995, Videocon launched an ad campaign to further its standing in the Indian electronics industry by promoting its wide range of consumer electronics such as televisions, VCRs, washing machines, refrigerators, air conditioners, and audio systems. While the headline proudly declared that Videocon is "Exploring the most advanced technologies to give you the world's finest Consumer Electronics and Home Appliances," the slogan at the bottom of the ad invited readers to "Bring Home the Leader." The subheading below the headline, also highlighted in a bold typeface, proudly announces: "Videocon. A household name." Two other subheadings, listed above the slogan— "Bring Home the Leader"—declare that Videocon is "Bringing the world to India" and "Improving the quality of life." When read in relation to text under each of the three subheadings in the ad, Videocon's slogan—"Bring Home the Leader"—appears to be an ambivalent dual invitation to the reader (see figure 8).[62]

In one sense, the slogan invites the reader at home to buy Videocon's range of products, since the company is, as the ad copy declares, "the leader in Consumer Electronics and Home Appliances with 19 state-of-the-art manufacturing facilities all over India." In another sense, the slogan "Bring Home the Leader" appears to invite the reader to embrace a new nationalist strategy of bringing into India "a wide range of quality products" made by Videocon in collaboration with world leaders in the electronics industry. The ad copy informs the reader that Videocon's color televisions, VCRs, and VCPs are manufactured "through a technical tie-up with Toshiba, Japan." The rest of the ad copy describes Videocon's technical tie-ups with Matsushita, Japan, for its washing machines, refrigerators, and air conditioners, and with Sansui, Japan, for its audio systems. The final section of the ad copy waxes eloquent on Videocon's corporate philosophy of "Improving the quality of life" by "bringing home the benefits of modern technological innovations to more and more people."[63]

Videocon. Exploring the most advanced technologies to give you the world's finest Consumer Electronics and Home Appliances.

Videocon. A household name.

Videocon is today the leader in Consumer Electronics and Home Appliances, with 19 state-of-the-art manufacturing facilities all over India, manned by skilled, dedicated employees.

* Washing Machines in technical collaboration with Matsushita Electric Industrial Co., Japan.

* Refrigerators, manufactured in technical design and drawing collaboration with Matsushita Refrigeration Co., Japan, manufacturer of the National Refrigerator.

* Air-conditioners manufactured under a design & drawing agreement with Matsushita Electric Industrial Co., Japan, owners of the brand 'National'.

* Audio Systems from Sansui, Japan.

The 32 Double Window Dramatic Wide Colour TV

Bringing the world to India

Videocon produces a wide range of quality products through its tie-ups with world leaders:

* Colour Televisions, VCRs and VCPs through a technical tie-up with Toshiba, Japan.

Improving the quality of life

The growing world of Videocon. A world led by technology. And a philosophy aimed at bringing home the benefits of modern technological innovations to more and more people.

VIDEOCON
BRING HOME THE LEADER

IB&W/VIL/DS/95/096

Figure 8. Videocon advertisement. *India Today,* February 29, 1996, 85.

In the center of the ad is the image of a television set with a caption underneath identifying the model as "The 32 Double Window Dramatic Wide Colour TV." The obvious reference here is to Videocon's innovative split-screen technology that enables viewers at home to see two programs at once on their television sets. The screen on the television set is equally divided into two windows. On the left side of the screen is the image of a game of cricket; on the right is an image of the Taj Mahal. The two images on the television screen in the Videocon ad also capture the ambivalent fears and fantasies of imagining the nation as a community in the world of electronic capitalism. On the one hand, the split-screen effect of the ad can be read as the nationalistic ambivalence toward schizophrenic fragmentation of the social reality by technological innovations that consumer electronics and home appliances bring into households across India. On the other hand, it also reveals the ambiguous potential of tie-ups between postcolonial nationalism and electronic capitalism that Videocon represents through its promise of "Bringing the world to India."[64]

For instance, the image on the left of the Videocon-Toshiba television screen—the game of cricket—depicts the ambivalent pleasures of assimilating the postcolonial into the world of the erstwhile colonizers by playing on the patriotic desire to witness a nationalist triumph in the once imperial pastime. The image on the right of the screen, the Taj Mahal, represents, with all its Orientalist ambiguity, an essentialized identity that is exclusively "Indian." Never mind the technological reification, the ad hails the readers; thanks to the Videocon-Toshiba "Double Window Dramatic Wide Colour TV," now Indians can at once partake in the world of electronic capitalism *and* retain their national identity, while being ensconced in the comfort of their own homes.

While Videocon celebrated its "tie-ups with world leaders" in the electronics industry as a corporate philosophy for "improving the quality of life" in India, BPL pursued an alternative strategy of recasting its internationalist ambitions in nationalist terms to capture the imagination of the domestic markets. "India's Pride, Britisher's Prize," proclaimed an advertisement for BPL color televisions.[65] In an obvious attempt to tug at the patriotic chords of the Indian consumer, the ad hails the reader to take pride in a domestic company that not only triumphs over competing foreign brands at home but also accomplishes a hitherto unheard-of feat of bringing international acclaim for the national community in the fiercely competitive world of electronic capitalism. As the ad copy elaborates:

BPL Color Televisions. The pride of millions in India. Now exported to Britain. And even awarded the "Best TV" in Britain by the prestigious "What video" magazine. After gaining approval from the British testhouse for compliance with British standards. BPL is the only Indian Company to have exported over 25,000 sets to Britain, in just a few months. Today BPL is all set to double this export sale from 5,000 to 10,000 per month. Go for the internationally proven BPL quality in India too. The pride of Indians. That's prized by the British.[66]

The creation of brand identification through invocations of nationalist pride and internationalist ambitions is central to the corporate philosophy of BPL. Anand Narsimha, the head of corporate brand management at BPL, admits that the "BPL brand name is worth more to the company than anything else that figures on or off the balance sheet."[67] In a survey conducted in 1997, BPL was ranked as "the most admired marketing company" in the consumer durables segment of the Indian market and listed as the sixth most popular brand overall.[68]

Onida—which had established high brand recognition in the 1980s with its slogan "Neighbour's envy. Owner's pride"—also fought hard to stay in the reckoning by recasting its popular campaign featuring the devil as a mascot, with a new nationalist slogan, "World's Envy. India's Pride."[69] The change of slogan in the 1990s clearly represents an acknowledgment by Onida that television envy is longer a domestic affair between neighbors within the national community but rather a transnational phenomenon where Indians now have to compete on the global stage of electronic capitalism.

"Ominous signs speak of unquenched thirst for global competition," declares the headline in this ad for Onida's state-of-the-art KY Series color television set.[70] With dark clouds looming in the background, the ad imagery depicts Onida's mascot, the devil, staring down from the skies above the Eiffel Tower in Paris, which is swaying precariously after being snapped from its moorings on one side. The following text provides a glimpse into what the devil Onida is up to in this advertisement:

> The hour has come. Face the chilling truth. The thirst for competition has reached raging new heights.
>
> You can hear the signs. The exclusive G-Horn Theatre Sound with Power Bass makes others quietly tremble.
>
> You can see the signs. An exclusive high resolution, black tinted, Flat Screen that creates perfect colours, sharpest pictures, making others go pale with fear.

You can control the signs. The super-intelligent Master Command with Menu-driven Full Function Remote gives you the power to control every aspect of the most envied TV in the world.

Indeed, the signs are ominous. For neighbours who may have settled for less.[71]

The "unquenched thirst for global competition" that the Onida ad speaks of as an "ominous sign" of things to come was, in fact, a pressing concern for all the major domestic companies in the electronics industry after the government of India announced new policies for further deregulation of the television manufacturing business in November 1996. As part of its deregulatory policies, the government decreased the import duties on color picture tubes and allowed for greater competition between foreign corporations and domestic companies in the television manufacturing business.

Between 1998 and 1999, the color television industry in India witnessed a 30 percent growth rate, due in large part to the successful crop yields for many farmers in the agricultural sector, the announcement of pay raises for all employees in the government sector, and the enthusiastic response of viewers across the nation to the World Cup Cricket matches in 1999. During the period when the World Cup matches were being telecast in India, there were reports of sales of color television sets increasing by a phenomenal 40 to 50 percent. L.G., the official sponsor of the World Cup, strategically used the popularity of the event to promote its brand around the country, and the company witnessed a staggering 95 percent growth in 1999.[72] "Do you see a revolution here?" asked the headline for an advertisement for L.G. Flatron monitors and color television sets.[73] With the close-up of an apple on a large flat screen on one side, the ad copy on the other side explains: "Many a [*sic*] people have seen apples fall down. But when Newton saw it, he saw gravity in it. A discovery that set the Space Revolution rolling. Just that added passion to see different. And a revolution is on hand."[74]

The revolution that the L.G. advertisement anticipated had little to do with Newtonian revelations about a falling apple and more to do with the visual representation of that apple on flat-screen monitors, such as L.G.'s Flatron television, which had become a rage in the Indian electronics industry. Industry experts estimate that flat-screen technology is the fastest growing segment of the market, and it was expected to account for 10 percent of the color television sets sold in India by 2004. With every major domestic and foreign company in the Indian electronics industry eyeing this new segment in the television manufacturing business, the end of the 1990s witnessed a spate of advertisements promoting the latest in flat-screen and surround-sound technologies.

Since the transnational brand names like L.G., Samsung, and Sony are not identified as "Indian" by the consumers or by the corporations themselves, the advertisements often celebrated the foreignness of their components by emphasizing the higher quality of their imported state-of-the-art technologies. Therefore, unlike the ads for television sets made by Indian manufacturers such as BPL, Onida, and Videocon in the early 1990s, the ads for flat-screen monitors introduced by the transnational corporations in the late 1990s do not claim a sense of "authenticity" by tugging at the nationalist sentiments of the consumers.

Rather, by highlighting the innovative and high-quality foreign components, the ads for the transnational brands argue that their flat-screen television sets provide a more authentic representation of the outside world than the curved-screen monitors that most Indian viewers use in their homes. Yet, at the same time, many of the foreign brands were careful not to ignore the necessity of customizing their television sets to suit the domestic needs of the Indian consumer—particularly in the low-end and middle-range models.

L.G. Electronics of Korea, which made two unsuccessful bids to capture the Indian market in the early to mid-1990s, embraced a new strategy of "Indianizing" its products by unveiling its much-publicized "Sampoorna" color television sets in 1997. Aimed at Hindi-speaking consumers in the semiurban and rural areas of North India, Sampoorna is the first color television set capable of displaying text in Devanagari script on the screen. However, in a comprehensive survey of the color television industry, India Infoline found that the marketing strategies used by television manufacturers like L.G. to create product differentiation for urban, semiurban, and rural consumers went completely awry in the late 1990s. As the India Infoline report puts it: "All these years, it was felt that the rural markets could not afford premium products, but models specially designed for the rural markets have found more takers in the urban markets while premium models have sold more in rural areas."[75]

For instance, the India Infoline report found that L.G.'s state-of-the-art Flatron monitors and Onida's KY Series color television sets with "G-Horn Theatre Sound" were more popular among rural and semiurban consumers, while the Sampoorna model created for the rural markets by L.G. found more buyers in urban areas. One of the reasons is that the rural rich see television sets as a long-term investment and are therefore more willing to consider premium high-end models as replacements for their old black-and-white television sets. Meanwhile, the urban middle-class consumer is more interested in utilizing the long-term financing options and the price-slashing discount schemes on the low- to middle-range models provided by both manufacturers and retailers who want to make room on their shelves for the newest products in the market.

Sony, which aims its latest innovations at the high end of the television market, launched its Wega series with three 21–inch TVs and a top-of-the-line home theater system in May 1999. Sony's television sets are priced 20–25 percent higher than other brands because the company imports its picture tubes, and its television sets have very few components that are indigenously produced in India. Therefore, the advertising campaigns for Sony's Wega television sets also showed a shift in emphasis from the strategy used by other television manufacturers, who explicitly link their brand names and corporate identities to nationalist imaginations, as in the aforementioned ad for L.G.'s Sampoorna TV.

"It takes the tele out of television," Sony proclaimed in the headline of an ad campaign promoting its WEGA DRC flat-screen technology, "with 4 times the normal picture density" available on other television monitors.[76] The image under the headline shows a couple sitting together in front of an oversized flat screen that looks more like a large cinema screen. On the screen is an extreme close-up of a ferocious-looking shark with its jaws wide open, as if ready to devour the couple sitting in front of it (see figure 9).

Another variant of the Sony Wega ad depicts the oversized flat screen, this time depicting a high-speed bicycling race.[77] Once again, simulating a sense of high-speed motion, the lead cyclist on the screen appears to almost run over the couple watching the sporting event. After providing this tantalizing glimpse into the viewing experience of Sony WEGA monitors, the two ads end with the following invitation to the reader: "Welcome to the real world." The text at the bottom of the page is identical in both the Sony television ads and presents in greater detail the considerable technological innovations of the WEGA DRC flat-screen monitors:

Presenting the WEGA DRC with 4 times the normal picture density.

    To take you 4 times closer to the action. 4 times closer to reality. That's the magic of Digital Reality Creation, Sony's unique digital-signal processing technology. It doubles the vertical and horizontal density, eliminates all visible scanning lines and removes flicker. Even text remains super still, giving you a high definition visual experience that's about 4 times better, richer, and so up close, it couldn't be further from television.[78]

Figure 9. Advertisement for Sony high-definition televisions. *India Today,* May 21, 2001.

The objective of the Sony Wega ad appears to be crystal clear: to overwhelm the reader with as many technological innovations as possible, like "Dual Exhaust 3–D System," "DVD Component Input Terminal," "Vertical Compression Technology," and "Eco Mode," in order to provide the promise of "a high definition visual experience" of the world of television that, paradoxically enough, "couldn't be further from television."[79] Samsung, which entered the Indian markets in December 1995, also released an ad campaign to promote its version of "television reality" on a flat-screen monitor with its trademarked DynaFlat digital technology. "Amazingly life-like" was the slogan that Samsung embraced in proclaiming the technological virtues of its Plano Digital Flat TV monitors.[80]

The ad imagery makes the point abundantly clear by showing a family in their living room hiding behind a bunker of sandbags and a barbed-wire fence that separates the couch from the television set with an image of two military men standing next to a machine gun. The headline above their heads announces the television schedule that would enable the family members to turn their living room into a war room: "8:00 P.M. 'GUNS OF NAVARONE' ON SAMSUNG PLANO, THE DIGITAL FLAT TV" (see figure 10).[81]

Another variant of the Samsung Plano ad features the family members sitting on a couch in their living room and holding umbrellas over their heads as they watch a scene from a classic Hindi film, *Barsat* (Rain) on their flat-screen television set.[82] The headline announces this family appointment with television as "10:30 P.M. 'BARSAT.' ON SAMSUNG PLANO, THE DIGITAL FLAT TV." In both versions of the Samsung ad, the slogan at the bottom of the page reminds the reader of the "amazingly life like" experience of viewing films on a flat screen at home, instead of going out to the cinema theater. The ad copy elaborates: "Unlike other flat screen TVs that have curved images, Samsung Plano is perhaps the only flat screen television that shows flat images. The result of Samsung's unique DynaFlat Technology. Which is why, reality will hit you like nothing else has. Be careful of what you watch on Samsung Plano."[83]

In many ways, the Samsung and Sony ad campaigns at the end of the 1990s recall an old advertising strategy used by Televista TV in the 1970s to celebrate the viewing practices of television by promising the viewer at home a close-up view of the performance in public venues like cinema theaters. However, in the Televista campaign of the early 1970s, the ad imagery attempts to conceal the technological reification of television through exaggerated claims about the small screen's ability to bring home the cinematic screen by mirroring its realistic representations of the outside world.

Is There an *Indian Community of Television?*

By the end of the 1990s, what we witness in the Samsung and Sony ad campaigns is a complete inversion of the early promise of the Televista ads to purely reflect the cinematic screen on the television screen and a bold new invitation to immerse oneself and one's family in the electronic frontier of the "home theater," where "reality will hit you like nothing else has." If the new technologies of high-fidelity sound and image were used in the early Televista advertisements for mirroring the outside world inside the television set, then the Samsung and Sony campaigns invoke newer technologies of surround-sound and flat-screen monitors to shatter that mirror of reality in the home theater. For instance, the Samsung ad campaign seeks to emphasize the technological innovations of television in great detail to convince the reader of the home theater's ability to create "an amazingly life-like" experience of the outside world inside the living room. Outlining the sophisticated technology in Plano Digital Flat TV, the Samsung ad copy at the bottom of the page explains:

> DynaFlat. Optimum curvature of the internal surface in the DynaFlat CRT helps realise perfect flat images without the concaveness of conventional flat CRTs.
>   Impact PORT Speakers. The Impact PORT adopts superior aerodynamics to maximize true sound. The result is a clearer high impact bass reproduction.
>   Super Pigment Phosphor. High brightness phosphors deliver vivid, life-like images.
>   Dolby Surround Pro Logic. This feature delivers ear-pleasing 360-degree surround sound. It is also equipped with external audio outputs for the total home theatre experience.[84]

What is most striking about the ad campaigns for state-of-the-art television technologies launched by Sony and Samsung in the Indian markets is the complete inversion of the principles of realism, which is a characteristic feature of imagination in modern societies. Therefore, I argue that as the Samsung and Sony campaigns promise to take "the tele out of television," in their exaggerated claims about the abilities of the home theater to create "amazingly life like" images, they invoke the discourse of hyperrealism that is considered a characteristic feature of electronic capitalism. In the hyperreal world of the home theater, the relationship between the television screen

Figure 10. Advertisement for Samsung Digital Plano Flat TV. *India Today,* April 2, 2001, 22–23.

and the cinematic screen is no longer defined through the promise of providing a mirror image of the outside world.

Mirroring the outside world is no longer adequate, for what television provides the consumer through its televiewing practices of *door-darshan* is a world of its own—with a plethora of domestic and foreign channels available to the viewer at the click of a remote control. In effect, television now provides a total lifestyle for the willing consumer of electronic capitalism. What is inside the television set, the advertisements promise, *is* a reality in and by itself that no longer appears to need an outside reality for legitimation.

As traditional distinctions of the outside and the inside, the foreign and the domestic, or the home and the world are blurred by the rapid advances of electronic capitalism, television viewers are induced to imagine the multitude of real-life events represented on the screen much as the "reel life" of cinematic events. The defining element of these on-screen events is the cut, or an imaginary rupture, which, as Arjun Appadurai reminds us, creates a disjuncture between cinematic or televisual frames even as it erases their differences to construct a seamless narrative that appears more real than the off-screen reality it seeks to represent in the home theater.[85] "Which is why," as the Samsung ads caution the reader, "reality will hit you like nothing else has."

Gandhian glasses—the iconic symbols of the nationalist vision in India—are no longer adequate to survey the viewing practices of *door-darshan* on television. "Seeing bifocally," to borrow John Durham Peters's elegant formulation, becomes an essential ingredient of viewing television in the home theaters of electronic capitalism.[86] Commenting on the role of new media in changing the relationship between the inside and the outside, the domestic and the foreign, or the global and the local, Peters writes:

> Our vision of the social world is bifocal. Institutions of the global constitute totalities that we could otherwise experience only in pieces, such as populations, the weather, employment, inflation, the gross national product, or public opinion. The irony is that the general becomes clear through representation, whereas the immediate is subject to the fragmenting effects of our limited experience. Our sense organs, having evolved over the ages to capture immediate experience of the local, find themselves cheated of their prey.[87]

The bifocal vision of electronic media that Peters refers to should be seen not as an extraneous influence on the work of cultural imagination in society but as a fundamental ingredient of the interrogation, subversion, and transformation of the sense of imagination. As Appadurai argues, the technologies of television are now pervasive in society, "through such effects as the telescoping of news into audio-video bytes, through the tension between the public spaces of cinema and the more exclusive spaces of video watching, through the immediacy of their absorption into public discourse, and through their tendency to be associated with glamour, cosmopolitanism, and the new."[88]

Without getting carried away by the unimaginable possibilities of losing ourselves in the hyperreal excesses of home theaters, or without being troubled by the social perils of living in a topsy-turvy world of electronic capitalism, it is important to remember that such utopian desires and dystopian fears have always been integral to cultural imaginations of new media in all societies. As Spigel points out, given the electronic media's ability to bring the outside world into the home, it is not surprising that television has been often depicted as "the ultimate expression of progress in utopian statements concerning 'man's' ability to conquer and to domesticate space."[89] She finds that the concept of "home theater" has been central to debates about the role of electronic media in society at least since the turn of the twentieth century. Citing an article published in an American magazine, the *Independent,* in 1912 that touts the arrival of "The Future Home Theater," Spigel argues:

> the magazine promised such dreams would come true through the develop-
> ment and application of two technologies. Sound and image could be trans-
> mitted through telephone wires "instantaneously from a central stage" or
> recorded through a combination of film and disk ("talking pictures"), which
> in turn might be sent through the telephone wires. In case these elaborate
> plans seemed excessively strange for the home environment, the magazine
> promised that the new "electric theater . . . will not seem a mechanical device,
> but a window or a pair of magic opera glasses through which one will watch
> the actors or the doers." This window would open onto "vistas of reality,"
> illusions far better than the "flat, flickering, black and white [motion] pictures
> of today," illusions produced through a combination of color, music, and
> 3–D photography. Best of all, the magazine predicted that these "inventions
> will become cheap enough to be, like the country telephone, in every room,
> so that one can go to the theater without leaving the sitting room."[90]

What is now being promoted as the viewing practices of television in commercials for the home theaters of the twenty-first century is, therefore, not anything radically new in terms of the creative ways in which techno-logical enthusiasts have always imagined the role of electronic media in so-ciety. This is not to suggest that the work of cultural imagination in elec-tronic capitalism is merely the repetition of history. Rather, it is to recognize that "the discursive conventions for thinking about communication tech-nologies are very much the same,"[91] while the historical contexts in which they operate change considerably.

As I have shown, in the case of television advertising in India from the beginning of the 1970s to the end of the 1990s, the cultural themes of hyper-reality invoked to sell the latest innovations in flat-screen technologies in the year 2001 are not unlike those used by national companies such as Videocon, BPL, and Onida in advertisements for state-of-the-art color television sets in the early 1990s, or those used in the advertisements for black-and-white television sets in the 1970s; all of which in turn are based on the popular strategies used by technological enthusiasts and science fiction writers in years past. However, the historical contexts from which these cultural discourses emerge provide illuminating insights into the changing relationship between televisual imaginations—which I have defined in terms of the viewing practices of *door-darshan*—and the synthetic sense of reality that new electronic technologies engender through their constant mediation of the inside and the outside, the domestic and the foreign, or the home and the world in postcolonial India.

3 Between Tradition and Modernity: The Development of an "Indian" Community of Television

IN THIS CHAPTER, I address the question "Is there an *Indian* community of television?" by critically evaluating the utopian vision of using satellite communications for national development in the postcolonial world. In the first section, I outline the utopian theory of "development" and discuss how its central tenet—that satellite television can alleviate social underdevelopment—became the dominant paradigm in international communications in the second half of the twentieth century. In the second section, I discuss critiques of the "development" paradigm in terms of theories of "imperialism" that have been forcefully articulated by Marxist critics of Western capitalism; particularly by the advocates of "anticapitalist development" in the erstwhile Soviet Union. In the third section, I examine the "mixed" theories of development articulated by Indian nationalists such as Gandhi and Nehru in their attempts to imagine a postcolonial alternative to both Western capitalism and Soviet communism. In the final section, I conclude that the question of an *Indian* community of television must be addressed in terms of the postcolonial quest to imagine an "uncolonized" community of nationalism in the context of the ideological conflicts between capitalist models and noncapitalist modes of development in the Cold War and post–Cold War eras.

Television and (Inter)national Development

In the second half of the twentieth century, the rapid advance of satellite communications has been the harbinger of a new order of international communications whose most potent metaphor has been the notion of "development." It first gained official sanction in President Harry Truman's speech on January 20, 1949, when he pledged to mobilize his country's considerable technical skills and scientific knowledge to fight "underdevelopment" in the poorer nations of the world. In the years following World War II, Truman and his foreign policy advisors deemed it the responsibility of the United States to aid the underdeveloped nations of the world that did not possess the scientific knowledge and technical skills needed for the rapid development of their national communities and media systems.

Although significant in its implications for national development and international communications, the fight against "underdevelopment" as a cornerstone of U.S. foreign policy was included in Truman's speech almost as an afterthought. At the last minute, a civil servant had suggested adding a fourth point—about the U.S. government's plans to provide technical aid to the "underdeveloped" nations of the world—in addition to the three points the president and his speech writers where working on; namely, the United States support for the United Nations, the U.S. role in European reconstruction through the Marshall Plan, and the creation of NATO as a Euro-American defense system to counter the military threat of the Soviet Union. The next day's newspaper headlines were all about "Point Four," and neither the president nor the secretary of state in the United States government were able to provide any more details about it.[1] The only details that journalists could uncover were in the full text of "Point Four."[2]

At first glance, there is nothing remarkable about "Point Four" in Truman's speech, which is full of good political intentions but makes no economic commitments to allocate the resources necessary to launch this "bold new program" in the international arena. In fact, even as Truman says that the United States "is pre-eminent among nations in the development of industrial and scientific techniques," he cautions that the "material resources we can afford to use for assistance of other peoples is limited."[3] However, what is remarkable about this text is that in it the United States government places itself in charge of international affairs—with a cursory nod to the United Nations—by simply talking about mobilizing its "growing and inexhaustible" resources in "scientific and technical knowledge" to help the underdeveloped nations of the world.[4]

In doing so, Truman creates, what Gilbert Rist calls, a "a minor masterpiece" that elevates the American ideology of capitalist development to a geopolitical ideal that the "underdeveloped" parts of the world must work toward in their quest to attain peace and prosperity in their national communities. Therefore,

Truman's introduction of the term "underdevelopment" to define the newly independent nations "radically altered the way the world was seen."[5]

Truman, however, was not the first to introduce the term "underdevelopment" into international affairs. Wilfred Benson, a former member of the Secretariat of the International Labor Organization, used the term in 1942 when he wrote about the need to pay attention to the "underdeveloped areas" in order to achieve economic prosperity and peace in the world. However, Benson's formulation did not find much resonance at the time, though the term "underdevelopment" was occasionally invoked through the 1940s by experts in public policy and in United Nations documents. The notion of underdevelopment acquired significance only after Truman presented it as emblematic of his administration's foreign policy, forever changing the meaning of "development" in international affairs. As Gustavo Esteva puts it:

> Never before had a word been universally accepted on the very day of its political coinage. A new perception of one's own self, and of the other, was suddenly created. Two hundred years of social construction of the historical political meaning of the term, development, were successfully usurped and transmogrified. A political and philosophical proposition of Marx, packaged American-style as a struggle against communism and at the service of the hegemonic design of the United States, succeeded in permeating both the popular and intellectual mind for the rest of the century.[6]

Before the emergence of the ideological category called "underdevelopment," world affairs in Asia, Africa, and Latin America had been mostly articulated in terms of the colonizer/colonized relationship. By invoking the term "underdevelopment" to describe the postcolonial world, Truman proposed to dispense with "the old imperialism" based on "exploitation for foreign profit" and establish a new international order based on a community of nations in which "every state was equal *de jure,* even if it was not (yet) *de facto.*"[7] Drawing on the principles enshrined in the United Nations' Universal Declaration of Human Rights, the new ideology of international development projected a utopian ideal for "the progressive globalization of the system of States."[8]

The ideology of international development through greater cooperation between the developed and the underdeveloped nations also derives its legitimacy from an earlier model, known as the Marshall Plan, that was successfully implemented among the war-ravaged nations of Western Europe. In the euphoric aftermath of World War II, the victorious Allies, led by President Truman, extolled the virtues of cooperation between the United States and Western Europe in the triumph over the Axis powers. To further the Euro-American relationship in the postwar era, the United States unveiled the Marshall Plan in 1947. It was celebrated as a new foundation for Euro-American friendship, and the United States proclaimed its intention to aid the political, economic, and cultural development of Western Europe without any apparent gains for itself. At the heart of the Marshall Plan is the ideology of Western liberalism, which posits that free enterprise (capitalism plus democracy) leads to international development and, by extension, secures peace and prosperity for all.

After a hard-fought battle against fascism and its persuasive propaganda in Europe during World War II, politicians and policy-makers in the United States were convinced that Soviet communism and its powerful ideology, if unchecked, had the potential to take over the underdeveloped parts of the world. In 1950, Senator Karl E. Mundt of South Dakota proposed what he called "a bold plan" for the development of a global television network that "can be put to work overseas, as America's mightiest weapon in the battle against Communism's other allies, hunger and ignorance."[9] Calling it "The Vision of America," Mundt elaborated on his plan to bombard the world "with new facts and new ideas" as follows: "Programs would originate in each country, using both film and live actors. By utilizing local talent, speaking local dialects, programs could be geared directly to the interests and needs of the people."[10]

According to Mundt's view of international communications, underdevelopment is the root cause of the scarcity that leads to social unrest in the world. This unrest, when not controlled, leads to the disruption of the existing world order. The underdeveloped world, goes the argument, is particularly prone to disorder, since gullible natives who lack adequate scientific knowledge and technical skills are easily swayed by disinformation, and this could open the door to a possible takeover by despotic fascists or opportunistic communists.

Another plan, proposed in 1953 by Senator Alexander Wiley, the chairman of the Senate Foreign Relations Committee, called on the U.S. government to "take immediate steps in order to develop methods of transmitting television images across the seas."[11] Lead by Senators Karl Mundt and Bourke Hickenlooper, the U.S. Senate passed a joint resolution on July 23, 1953, authorizing the creation of a public and governmental commission to examine the possibility of developing an intercontinental telecommunications system that would allow the United States to use television for propagating democracy around the world. Advocating the creation of a Global Television Commission, Senator Hickenlooper referred to what he called "the great danger" of the Soviet Union jamming the telecommunication systems of "the free world." Explaining the reasons for his concerns, Senator Hickenlooper argued: "In the event of war, one of the first things that would happen would be for the enemy to do utmost to disrupt communications between the U.S. and Europe. It is technically possible now, however, to develop a system of microwave transmissions, from point to point across the North Atlantic, so that we may communicate not only TV images, but commercial and governmental messages."[12]

If President Truman's speech in 1949 paved the way for proposals to create a global television system using transoceanic cables and microwave relays, then it was President John F. Kennedy's speech in 1964—in which he declared that putting a man on the moon was one of America's national goals—that clearly provided a new momentum for the use of satellite communications to assert American supremacy in the Cold War era. Perhaps Kennedy's speech was the reaction of an astute politician who recognized the power of a growing nationalist desire to counter Russia's equation of the successful launch of Sputnik to a triumph of communism and its techniques of scientific socialism. Perhaps Kennedy was playing up the hype of the coming age of megascience in a technologically driven society.

The rhetorical significance of the speeches of the two American presidents is that they indicate the symbolic culmination of a two-hundred-year-old project of Euro-American liberalism: to legitimize the philosophical and ideological rationales of scientific rationality and technological progress as a universal ideal for human development by equating the practical payoffs of "scientific" technologies, such as satellite communications, with the most desirable traits of free market capitalism and liberal democracy.[13]

In the heady days following Kennedy's articulation of the nationalist ideology of the space race to the internationalist ideals of development, many Euro-American academics, journalists, and policy-makers embraced the policy of sharing the surplus of their scientific knowledge and technical resources with any "underdeveloped" nation willing to work toward modernization and, by an unsaid corollary, join the Cold War to combat Soviet communism. Soon, the Cold War struggles between the Euro-American nations (self-designated as the First World) and Soviet communism (defined as the Second World) were extended to the fight against underdevelopment in the newly independent postcolonial nations—now described as the Third World.[14]

In their book *World Peace via Satellite Communication* (1965), Herbert M. Frenkel and Richard E. Frenkel detailed an ambitious plan to realize the promise of President Kennedy's proposal to create "a global television system of communications satellites linking the whole world in telegraph, telephone, radio and television."[15] Claiming widespread support from American academics, activists, and politicians for the idea of promoting television as a medium for global peace, Frenkel and Frenkel cited the following newspaper article in the *New York Times* on October 11, 1961:

TV BY SATELLITE URGED FOR THE WORLD—MAGNUSON CITES IMPACT ON UNDER-DEVELOPED AREAS

Senator Warren G. Magnuson called upon the nation last night to speed the development of communication satellites able to carry educational television to the "uncommitted peoples and the under-developed areas of the world."[16]

Other American scholars of satellite communications looked fondly toward the Intelsat system for creating a global communications system based on Western principles of free market economics and democratic exchange of ideas.[17] Intelsat (The International Telecommunications Satellite Consortium) was created in 1964 to give its member nations access to a worldwide commercial satellite service. The structure and the charter of Intelsat as an international organization were established over the course of a decade; a first session of intense debates produced two interim agreements signed in 1964, and a second session of discussions led to the creation of two permanent agreements signed in 1971.[18]

From its inception, Intelsat was mired in political, economic, and regulatory debates because of the different priorities of the consortium members, particularly the United States and the Western European nations, which played a dominant role in the negotiations. However, these nations were united in their efforts to extend the reach of the Euro-American telecommunication system across the world and to use the advances in satellite technologies to curtail the global expansion of Soviet communism. Having experienced the persuasive power of propaganda during World War II, many Euro-American media experts placed immense faith in the ability of mass media like radio, television, and satellite communications to modernize traditional societies and sway public opinion in the underdeveloped world away from the ideological powers of communism.

Among the early proponents of modernization theory was Daniel Lerner, an expert in the study of psychological warfare through propaganda during World War II and the author of *Sykewar: Psychological Warfare against Germany.* In a later book, *The Passing of Traditional Society,* Lerner uses his expertise in psychological warfare to combat a new enemy of the Euro-American order—underdevelopment. In this book, subtitled *Modernizing the Middle East,* Lerner evaluated the role of electronic media in the modernization of "traditional" societies in Egypt, Jordan, Lebanon, Iran, Syria, and Turkey.[19]

Using survey results of 1,600 interviews conducted in these six countries, Lerner analyzes the effects on different sections of the population of listening to radio programs broadcast by major international broadcasters like Voice of America (VOA), BBC, and Radio Moscow. Drawing from these interviews, Lerner maps a correlation between development (equated with economic productivity) and modernization (defined in terms of a set of variables such as urbanization, mass media usage, literacy, democratic participation, etc.). In his findings, Lerner concludes that the introduction of electronic media into traditional societies (indicated by activities such as media usage and spread of literacy) leads to the development of a *mobile personality* among individuals who are highly *empathetic* to modernization.[20] Lerner defined *empathy* as the capacity of an individual to accept new aspects of modernity in a traditional society.

Advocating the need to harness the mobile personality of individuals who have empathy for modernization, Lerner proposes the use of his now-famous communications model to develop a *climate of acceptance* in traditional societies toward the Western ideology of free enterprise. According to this model, there are three distinct types of audiences in the underdeveloped world: (1) the "moderns," who are already in tune with the Euro-American audiences through their interest in international radio stations such as the VOA and the BBC; (2) the "traditionals," who are the most removed from, and thus most resistant to, the Euro-American ideals of modernization; and (3) the "transitionals," who can be persuaded to embrace the ideology of free enterprise when armed with scientific knowledge and technical resources by the Euro-American media expert.

What is most interesting, and perhaps more insidious, in Lerner's communication model for development through modernization is that the audience categories he uses in the Middle East are the same ones that he proposes in his earlier propaganda model for "Sykewar" against fascism in Nazi Germany. In his influential study of the power of propaganda during World War II in Germany, Lerner defines "moderns" as anti-Nazi (or pro-Allied) Germans, "traditionals" as the Nazi Germans, and "transitionals" as apolitical Germans who could be converted into allies through sustained counterpropaganda against Nazi disinformation. The parallels in the two studies are striking, and their political significance is revealing, given the ideological elevation of modernization through Euro-American notions of free enterprise as "development," and the underlying equation of archaic traditions with fascist ideologies of regress and "underdevelopment."[21]

When the propaganda model of "Sykewar" against fascism is extended to the communications model for the international war against underdevelopment, the "modern" audiences in the underdeveloped world, despite their geographic location (like antifascists in Nazi Germany), are considered to be the allies of—even if not on a par with—the "moderns" in Europe and the United States. Conversely, "traditional" audiences in the underdeveloped world (akin to the fascists in Nazi Germany) are the enemies of the Euro-American project of international development through modernization.

In other words, audiences who embrace nonmodern traditions are condemned to an illegitimate category of underdevelopment, much like the fascists in Nazi Germany who were ranged against the Western allies during World War II. The nonmodern traditions of the underdeveloped world are deemed legitimate only if they do not inhibit the Euro-American project of development, or if they can be enlisted to aid its international progress.

A significant group of audiences located between the legitimate "moderns" and the illegitimate "traditionals" is the hybrid category of "transitionals"—an audience group characterized by its ambivalence toward both nonmodern traditions and modern development—much like the ambivalent German audiences who were caught between the ideologies of Nazi fascism and Western liberalism. In Lerner's ambitious project for the modernization of traditional societies, the "transitionals" are neither "developed" nor "underdeveloped." Placed in a hybrid category of the "developing," transitional audiences are deemed to be potential allies of the Euro-American goals of development—if and when they are armed with the scientific knowledge and technical skills needed for modernization.

On the other hand, as Lerner's propaganda and communication models posit, in the absence of such knowledge, "transitionals" are also the audiences who could be swayed by disinformation and propaganda (of fascism during World War II against Germany, of communism in the Cold War against the Soviet Union, and of nonmodern traditions in the international war against underdevelopment).

Following Lerner's compelling schematic for conducting a "sykewar" against underdevelopment in the so-called Third World, there have been many attempts to apply, extend, and revise his communications model in order to refine the process of the transfer of scientific knowledge and technical skills from modern experts to the traditional laity using the principles of free enterprise. Among the earliest and most influential of such attempts has been the work of Everett M. Rogers, whose book *The Diffusion of In-*

novations quickly attained canonical stature of its own. In this book, Rogers also embraces the model of international development by advocating the rapid modernization of traditional societies in the underdeveloped and the developing worlds. Rogers argues for the use of mass media for the diffusion of scientific knowledge about innovative technologies, which would help modernize traditional societies in all areas of life, from agriculture, animal husbandry, and health and family welfare to housing, education, and literacy.[22]

In the following decades, under the watchful eyes of development experts, ambitious projects were launched in the quest to modernize audiences and mass media institutions in the underdeveloped and the developing nations of the so-called Third World. The proclaimed goal of these projects has been to develop modern media institutions and programming practices to persuade Third World audiences—particularly the "transitionals" in Lerner's scheme of things—to embrace and propagate the ideals of modernization, and by extension, the ideology of free enterprise.[23]

Given that the stated intention of the Euro-American experts was to cultivate the "transitional" audiences first, not surprisingly, many of the early experiments for the dissemination of scientific knowledge and technical skills through television were conducted in "developing" countries such as India and Brazil, whose national communities were deemed to be in a transitional category between the "developed" and "underdeveloped" dichotomy of the modernization process.

Under the aegis of the United Nations Organization and its agencies like UNESCO, many of the so-called First World nations, led by the United States, contributed to ambitious international projects in the so-called Third World countries such as Brazil and India. Although Brazil and India were not explicitly allied in the fight against world communism with the United States and the Western European nations, they were the earliest adopters of sophisticated technologies like satellite communications for education, literacy, poverty alleviation, and other similar nationalist agendas of development, which handsomely contributed to the Euro-American ambitions for global supremacy in the Cold War.

Contesting the Ideologies of International Development

Among the most articulate critics of the Euro-American model of international development are the proponents of Marxist theories of capitalist imperialism. Although there are many traditions in Asia, Africa, the Americas, and Europe that constitute the vast corpus of Marxist theories of imperialism, their common point of concern is the ideological critique of capitalism, which, they argue, obfuscates the underlying contradictions of class inequities in society. In the overall framework of international relations in the twentieth century, the most influential exponent of imperialism theory has been Vladimir Lenin, the architect of the Russian Revolution of 1917, which led to the spread of communism around the world.

In *Imperialism: The Highest Stage of Capitalism,* Lenin essentially embraces Marx's critique of class inequities in the capitalist economies of western Europe and extends it to the international realm. In this subtle extension of Marxist critiques of the national political economy of capitalism to the international realm, Lenin accounts for why inequities of capitalist production in European nations did not impoverish labor and enslave the working class as Marx had predicted but had instead improved living conditions for workers and had even enhanced the social status of their communities and organizations, such as trade unions. Lenin recognizes that imperialism is the instrument by which European capitalism could shield its labor by exporting working-class poverty to far-flung colonies and importing the "surplus capital" generated by cheap colonial labor and resources.

Therefore, for Lenin, the development of inequities in the capitalist mode of production cannot be understood merely in terms of a national political economy where powerful elites exploit the working class. Instead, he argued, capitalist development occurs only when powerful ownership interests within a national political economy dominate the working class but export the exploitation of labor to other nations around the world. This exportation of exploitative labor from the national political economy to the international economy results in "the highest stage of capitalism," which Lenin defines as "imperialism."[24]

Following the demise of European colonialism, there were several liberation movements in the former colonies of Africa, Asia, and Latin America. In the period from 1940 to 1983, the number of independent nations in the world grew from 70, of which 30 were in Europe, to 169, of with 32 were in Europe.[25] During the period of decolonization, the communist leadership in the Soviet Union clearly recognized the enormous political significance of the end of colonialism.

"The liberation of former colonies and semi-colonies was a strong political and ideological blow to the capitalist system. . . . It has ceased to exist in the shape that it assumed in the 19th century and which extended into the first half of the 20th century," noted the Political Report of the Central Committee to the Twenty-seventh Party Congress of the Communist Party of the Soviet Union (CPSU), in 1986.[26] The political leadership in the Soviet Communist Party saw in this global transformation an opportunity to convince the postcolonial world to move beyond nationalism (which they had now attained) by joining the internationalist struggle against Western imperialism and fulfilling the Marxist goal of attaining "true" liberation in classless communism through noncapitalist development.

Since the Cold War battle between the Soviet Union and the United States was also a space race from the earliest days, Russian scientists took keen interest in developing satellite technologies for global communications. In June 1961, the president of the Soviet Academy of Sciences asserted that "the use of communications, and satellites, and satellites for relay services would revolutionize communications and television services."[27] On October 4, 1954, the Soviet Union shocked the world, and particularly the National Aeronautics and Space Administration (NASA) in the United States, by sending the first manmade satellite, called Sputnik, into space. The Soviet Union was also the first in the world to launch a domestic satellite system, called *Orbita*, in 1966. The early successes of the Sputnik experiments led the Soviet writer P. Donetsky to proclaim with great optimism: "Despite Major Difficulties, A World Television System is Possible." He summarized the attempts made by Soviet scientists toward creating a "Cosmovision" of global satellite television as follows.

> Soon after the Soviet Union had launched the world's first man-made Earth satellite on October 4, 1957, the Soviet scientist S. Katayev put forward the theory and design of a project for super-long-range television using man-made Earth satellites. . . .
>
> Soviet scientists succeeded in using a radio-television line to transmit back to Earth the image of the reverse side of the Moon. Space television—Cosmovision—was born on August 11, 1962.[28]

On August 5, 1968, the Union of Soviet Socialist Republics (USSR), Bulgaria, Cuba, Czechoslovakia, Hungary, Mongolia, Poland, and Romania joined to create an international satellite communications system called Intersputnik. The creation of Intersputnik was a clear signal that the Soviet Union was not willing to concede to the dominance of the commercial Intelsat system developed by the United States and its allies in 1964. However, in the same year the membership of Intelsat grew to sixty-three countries, and by the 1960s Intelsat was solidly ahead, with satellite services across the Atlantic, the Pacific, and the Indian Ocean. Commenting on the potential political advantages

to the Soviet Union of creating a noncapitalist system of satellite communications, such as Intersputnik, in contrast to the capitalist system of Intelsat,
Jonathan Galloway argues:

> In the first place, the only members of Intersputnik are states and the voting
> procedure is one state, one vote. This approach would obviously appeal to
> many countries, for it contrasts favorably with the image of Intelsat as be-
> ing dominated by a private, United States corporation. Secondly, the Pre-
> amble to the draft contemplates the establishment of a system which would
> provide direct broadcasting from satellites. Such a system could bypass the
> need to establish expensive earth stations to relay signals from satellites in
> space. . . . Intersputnik could greatly aid the less developed countries by
> enabling them to circumvent the need to establish a costly infrastructure
> for communications. Thirdly, the Intersputnik draft agreement allows for
> more than one international communications satellite system, whereas one
> of the main criticisms of Intelsat is that the signers of the Interim Arrange-
> ments committed themselves to a single international system.[29]

Although Intersputnik did not compete with Intelsat on a commercial
basis, it provided the Soviet Union with an ideological tool to demonstrate
many of the weaknesses in the capitalist approach used by the United States
and its allies and to expose what the Marxist-Leninists saw as the imperialist tendencies of the Western quest for hegemony in the international satellite system. In the wake of the Marxist-Leninist theorization of capitalism as
the highest stage of imperialism, critiques of global capitalist integration
have been articulated with varying degrees of success by many scholars of
international communications.

One of the better known theories of global capitalist integration is the
world systems theory developed by Immanuel Wallerstein in the mid-
1970s.[30] In Wallerstein's theory, the world system "is a single division of
labor, comprising multiple cultural systems, multiple political entities and
even different modes of surplus appropriation."[31] Wallerstein argues that
a single interdependent world system is not a creation of contemporary
global capitalism. Rather, he argues, it has existed since the sixteenth century, beginning with European expansionism and colonialism. Since its
inception, goes the Wallersteinian argument, this world system has imposed
a hierarchy in which labor is exploited by capitalist ownership, which seeks
to accumulate surplus. Over time, as this exploitation evolves, the economy
spreads to other regions and nations, creating a world economy in which
strong nations dominate weaker ones.

However, the world system envisioned by Wallerstein consists not merely of weak and strong nations. Rather, it constitutes a three-level system in which the strong nations make up the core, the weak nations make up the periphery, and the not-so-weak (or not-so-strong) ones make up the semiperiphery. The semiperiphery nations, in Wallerstein's world system, take up the role that middlemen perform in the relationship between ownership and labor in a capitalist economy; or the role that the middle class performs in the relationship between the elites and the lower classes in a nationalist economy. In performing this role of a "go-between," the semiperipheral nations, according to Wallerstein, reinforce the interdependency of the core and the periphery, even as they claim a relative autonomy and flexibility in the otherwise rigid world system.

In Wallerstein's notion of an interdependent world system, the historical domination of international communications by the "core" Euro-American nations escalates the inequities of earlier forms of imperialism, such as colonialism, by forcing "peripheral" nations of the postcolonial world into a culture of perpetual dependency. Therefore, Wallersteinian critics of global capitalist integration suggest that even the apparently benevolent gifts of surplus resources, and the seemingly egalitarian exchanges of technological expertise and media resources among the core, periphery, and semi-periphery, are ideologically implicated in the historical project of imperialism.

In many of the newly independent nations of Africa, Asia, and Latin America, some of the more forceful advocates of noncapitalist development, taking their cue from Marxist theories emanating from the Soviet Union, western Europe, and North America, deemed it legitimate and even necessary to resort to coercion, subversion, and military intervention in order to enlist the "traditional" audiences of the postcolonial world into the cause of communism. However, in the ideological transformations of Marxist theory from Marx and Lenin to Stalin and beyond in the Cold War era, the quest for the liberation of the working classes and the underdeveloped communities in the postcolonial world morphed into an obsessive desire for the rapid development of an international order of communism. In the process, the Marxist theorists and their theories of noncapitalist development attained an imperialistic bearing and an oppressive civilizing mission that, ironically enough, paralleled and even surpassed the internationalist ambitions of Western capitalism that they sought to address.

Ashis Nandy's astute critiques of the impact of Marx's theories in colonial and postcolonial India reveal that many Marxists deemed it their "moral duty" to rescue colonized subjects both from the ills of European colonialism and from the senility of their own cultural traditions.[32] The irony of the imperialist ambitions in Marx's civilizing mission and the ideological transformations of the Marxist theories of anticapitalist development in the colonial context were not entirely lost on the nationalist leadership of the postcolonial world. In many postcolonial nations like India, Egypt, and Yugoslavia, the nationalist leaders pledged to piously guard their hard-won independence from newer forms of imperialism—be they capitalist or Marxist. Despite enormous political-economic constraints, they remained steadfast in their refusal to align their nations wholesale with either Western models of capitalist development or with the Soviet models of noncapitalist development in the international arena. Instead, the postcolonial leadership attempted to imagine home-grown alternatives to both Western capitalism and Soviet communism.

The Postcolonial Vision of National Development

In the newly independent nations of the postcolonial world, critiques of political, economic, and cultural imperialism informed by Marxist and Freudian diagnoses of human and social development were cautionary tales for nationalist leaders and the cultural literati, who were extremely suspicious of their former colonizers. Many of these leaders who resolved to piously guard their hard-won national freedom were drawn as much to the Western models of international development through the principles of free enterprise as to the noncapitalist modes of development proposed by Marxist, Leninist, and Stalinist theorists of international communism.

Throughout the postcolonial world, nationalist leaders attempted to culturally synthesize a "mixed" vision of development that would accommodate their longing for a unique uncolonized nation and their quest to develop a modern nation-state. Some of the exemplary attempts to develop a uniquely postcolonial path toward national development are Kwame Nkrumah's "African Socialism," Jomo Kenyatta's "Mount Kenya," Nehru's "mixed economy" in India, and Sukarno's combination of nationalism and socialism in Indonesia.

What was common in these diverse postcolonial contexts was a desire among the nationalist elites to imagine an uncolonized nation while working through the many pulls and contradictions of international development engendered by the ideological conflicts between Western capitalism and Soviet communism. In *Toward Freedom*, Nehru clearly speaks of his desire to synthesize a nationalist vision for India by learning from the mistakes of the ideological conflicts of Soviet communism and Western liberalism in the international arena:

> I have long been drawn to socialism and communism, and Russia has appealed to me. Much in Soviet Russia I dislike—the ruthless suppression of all contrary opinion, the wholesale regimentation, the unnecessary violence (as I thought) in carrying out various policies. But there was no lack of violence and suppression in the capitalist world, and I realized more and more how the very basis and foundation of our acquisitive society and property was violence.[33]

As Gyan Prakash rightly argues, "the irresistible power of the urge for social justice and cultural renewal" in Nehru's "mixed vision" of nationalism is best captured in the notion of nonalignment, first formulated at the Bandung Conference, held in the Indonesian city of Bandung in April 1955.[34] The Bandung Conference was a major step toward the establishment of what has come to be known as the Non-Aligned Movement. The Bandung Conference was attended by delegates from twenty-nine countries—twenty-three from Asia and six from Africa. In a spirit of anticolonial, anti-imperialist solidarity, the participants of the Conference adopted a Declaration on the Promotion of Peace and Cooperation. In this Declaration, the signatory nations pledged to live together in peace as good neighborly nations and to develop their friendship and cooperation on the basis of the principles of mutual respect for sovereignty, territorial integrity, nonaggression, noninterference, and equality.[35]

Although the Bandung Conference brought to the forefront an alternative formation that envisioned a "third way" of international development in opposition to Western capitalism and Soviet communism, the irreducible differences within the Non-Aligned Movement were already in evidence. While the participant nations were united in their affirmation of nonalignment and their opposition to old and new forms of imperialism, the ambiguity and vagueness in Conference resolutions reveal that there was little agreement about what precisely these terms constitute.

Such was the diversity and disparity of viewpoints expressed by the participants of the Bandung Conference, that in the final resolution that outlined the principles of nonalignment, no mention was made of imperialism. However, despite its inevitable fallacies, the Bandung Conference Declaration remains a significant moment, in that it represented, for the first time, a bold attempt by the newly independent nations to imagine a postcolonial alternative to the international hegemony of Western capitalism and Soviet communism.

For Indian nationalists like Nehru, the Non-Aligned Movement provided the means to reconcile the irreducible difference between their nativist longing for an uncolonized nation and their more cosmopolitan concerns for an egalitarian world order. However, the creativity of the Nehruvian "mixed economy" lies in its ability to strategically integrate the nativist longings of postcolonial nationalism with Western-style market capitalism and Soviet-style, state-regulated socialism to provide a synthetic, hybrid, home-grown version of development that could claim to be self-reliant, even as it struggled to remain engaged in the international geopolitics of the Cold War.

When India gained its independence at the stroke of the midnight hour on August 15, 1947, jubilant Indian nationalists, led by Nehru, celebrated at the historic Red Fort in Delhi (the symbolic seat of power of the precolonial Mughal empire, and later the colonial British empire). In this moment of nationalist glory, even the anguish and pain of partitioning the Indian subcontinent into two nation-states (India and Pakistan) had been briefly forgotten, and the devastating communal riots between Hindus and Muslims that had engulfed the subcontinent in the wake of the partition were temporarily suspended; it was time to celebrate. As the Indian nationalists rejoiced at their independence from British colonialism, Mahatma Gandhi, the "Father of the Nation," was far away in a little village near Kolkata providing succor to strife-torn victims of communal riots.

Many in India have heard, read, and narrated this moving episode, which has been used by the Mahatma's admirers to proclaim his sainthood, material detachment, and human compassion, while his detractors see it as a marker of Gandhian eccentricity. However, given the intensity of emotions generated in this historic moment, I would argue that one must read Gandhi's response to Indian independence as much more than a symbolic gesture of a "spiritual saint" or as the woolly-headed sentimentalism of an eccentric old man. Instead, Gandhi's response to the birth of the modern Indian nation-state should be understood as a metaphor for his radically different vision of postcolonial nationalism.

For many Indian nationalists like Nehru, British colonialism had essentially ended with the departure of the colonizers and the arrival of the modern nation-state. Gandhi, for whom the battle against colonialism went beyond the legal establishment of a modern nation-state, found little to rejoice in, and continued his relentless fight against colonial violence and oppression, both internal and external. Many Indian nationalists were at once dismayed, embarrassed, and overawed by the seemingly detached indifference of the "Father of the Nation" toward the "realization" of a dream they had dreamt together. They respected the worldview he represented but could never quite understand it.

Here, it is important to note that the mainstream Indian nationalist movement under Gandhi was not a monolithic "Gandhian" faith. Some of Gandhi's greatest admirers and followers were staunch believers in Western ideals of nation-building and statecraft; others were ardent supporters of Marxist theories of anticapitalism and revolutionary communism; and there were many others who had equal admiration for Western notions of international development and Marxist ideals of classless communism (Nehru being a prime example). In this heterogeneous terrain of Gandhian nationalism, the Nehruvian strategy was to create a national policy of cultural synthesis. Nehru believed that the role of the modern nation-state was, as Prakash puts it, "to represent the myth-ridden people whose welfare and transformation . . . could not be left to the unrestrained play of capitalism."[36]

Thus, Prakash finds that the Indian nationalist elites led by Nehru could willingly "concede that Gandhi had the uncanny ability to read the pulse of the "irrational peasants," while at the same time brushing aside the Gandhian "reservations on modern industry and politics."[37] However, in brushing aside the Gandhian reservations, the Indian nationalist elites, led by Nehru, had in essence distanced themselves from the radically different vision of nationalism that Gandhi had provided in the anticolonial struggle. Despite the irreducible differences between the Gandhian and Nehruvian ideals, the nationalist leaders were able to synthesize a uniquely "Indian" response to colonialism that Partha Chatterjee has described as a "passive revolution." In *Nationalist Thought and the Colonial World,* Chatterjee uses this term to describe how Indian nationalists were able to gain the consent of the peasants, laborers, women, untouchables, and other subaltern groups, even though the leaders' elitist views of nation-building and statecraft were heavily influenced by the ideals and ideologies of Western liberalism and Soviet communism.[38] In this context, the "beauty of the Gandhian intervention" for the nationalist elites was that "it could deliver the popular forces," as Prakash puts it, "without ceding them the initiative."[39]

In an illuminating rereading of the Gandhian intervention in the Indian nationalist movement, Ashis Nandy ascribes Gandhi's phenomenal sway over the masses to his uncanny ability to speak and understand their everyday "language of survival," even while engaging in an ideological battle with the regimes of modern colonialism.[40] Anticipating much of the postmodern censure of modernity and its institutions, Gandhi formulated and practiced a simple theory of nonmodern living in a modern era, which he described as *swaraj* (self-rule) and *swadeshi* (self-reliance). That the Gandhian notions of *swaraj* and *swadeshi* suggest self-rule for the Indian nation as a community and self-reliance among the national subjects as individuals is no mere coincidence. As Ashis Nandy and Shiv Viswanath tell us, the underlying principle of the Gandhian theory of *swaraj* and *swadeshi* is the metaphor of the body and the homology between the politics of the body and the body politic.[41] Nandy and Viswanath write: "Bodily scale defines not only the nature of activity, but prescribes its limits. Modern mechanistic civilization is a disease because it violates the integrity of the body. The real tool . . . should be a natural extension of the body, not disjuctive with it. From these simple premises he [Gandhi] outlines a critique of modernity."[42]

Gandhi's critique of modern colonialism attempted to delegitimize the homology between the historical developmental cycle and human developmental cycle that enabled the colonizers to see Indian civilization as old and decrepit and, by extension, the colonized Indian subject as savage, infantile, and primitive. If modern developmental cycles in the Hegelian, Marxist, and Freudian traditions saw the present (modernity) as "a special case of an unfolding history," then for Gandhi "history was a special case of an all-embracing permanent present, waiting to be interpreted and reinterpreted."[43]

By firmly locating the "past" in the "present," Gandhi sought to delegitimize the historical developmental cycle of modern colonialism, which was heavily influenced by Hegelian/Marxist teleologies, Freudian myths of Oedipus, and Darwinian theories of social evolution. Nandy argues that at the level of the individual life cycle, Gandhi simultaneously delegitimized the discourse of modern development, thereby setting up a gender hierarchy in which the Western definitions of masculinity were considered superior to femininity, which was only better than hermaphroditism.

Gandhi perceived that a majority of communities in India were less influenced by the Western worldview than were most of the nationalist elites who led the freedom struggle against British colonialism. On the basis of this perception, Gandhi envisioned an alternative to Western developmental cycles by drawing from Hindu cosmology, which posited the androgyny of mythic gods such as Krishna and Shiva as superior to the human notion of masculinity and femininity. That is, "manliness and womanliness are equal, but the ability to overcome the man-woman dichotomy is superior to both, being an indicator of godly and saintly qualities."[44] The political implications of this reordering of gender were significant: if the developmental cycles of Western colonialism lumped femininity, loss of masculinity, and androgyny together and equated them with hermaphroditism, then Gandhi sought to delegitimize the ideology of colonialism by equating Western ideals of hypermasculinity with cowardice (a man who cannot accept his androgyny and femininity is a coward). By celebrating the mythology of androgyny in Hindu epics such as the *Mahabharata,* the *Ramayana,* and the *Shivapurana,* Gandhi sought to liberate both the colonizer and the colonized from what Ivan Illich has called the diagnostic imperialism of modern theories of development in the body politic and the individual body.[45]

Indeed, Gandhi was very selective in his strategic invocation of the Hindu epics to fit mainstream values of Indian society. At the same time, he was rather critical of non-Brahminical, or nonheterosexual, discussions of androgyny and gender relations in Hindu culture and Indian society. For instance, the orgiastic excesses and transgressive rituals practiced by tantric sects at the margins of Hinduism, or the scathing critiques of the Hindu caste hierarchy by populist radicals in the Indo-Arabic *sufi* movements, or the complete dismissal of Hindu theocracy by deconstructionists in Buddhist traditions and atheists in rationalist circles found little or no voice in Gandhi's vision of an "Indian" nation.

Moreover, Gandhi was certainly not the first postcolonial nationalist to recognize the homology between the body politic and the politics of the body in colonial discourse. Subhash Chandra Bose's revolutionary designs for the militaristic overthrow of British colonialism through strategic alliances with Germany and Japan and B. R. Ambedkar's clarion call for the democratic integration of untouchables and noncaste Indians into the political, economic, and cultural mainstream of the national community are two of the most prominent challenges to Gandhi's advocacy of *swaraj* and *swadeshi* in colonial and postcolonial India. Thus, the ideological distances among the Nehruvian agenda of state-sponsored development, the Gandhian notions of *swaraj* and *swadeshi,* Bose's revolutionary politics of anti-imperialism, and Ambedkar's critiques of the caste system, among others, has contributed to some of the most creative and often controversial debates over the quest for an uncolonized vision of nationalism in postcolonial India.

Is There an Indian *Community of Television?*

A wit among media circles once observed that while Jawaharlal Nehru was a visionary, his daughter Indira Gandhi was a televisionary. The quip, of course,

was a cruel, backhanded compliment to the authoritarian ways in which Indira Gandhi shrewdly manipulated Indian television to project her political
image by using the state-sponsored network, Doordarshan, for what her critics derisively called "Indira Darshan." Although sarcastic, this observation
does recognize that in matters of Indian television, Indira Gandhi, however
authoritarian, was indeed a visionary. Soon after she became prime minister
in 1966, she envisioned an ambitious plan for a national television network
in India. She commissioned her one-time classmate Vikram Sarabhai, of the
Indian Space Research Organization (ISRO), to work on the development of
television in India. In 1969, Sarabhai outlined a blueprint for creating a national television network within a decade. In an ambitious plan entitled "Television for Development," Sarabhai proclaimed: "A national program which
would provide television to about eighty per cent of India's population during the next ten years would be of great significance to national integration,
for implementing schemes of social and economic development, and for the
stimulation and promotion of the electronics industry. It is of particular significance to the large population living in isolated communities."[46]

To attain their nationalist goals of development, the postcolonial elites,
led by Indira Gandhi and Vikram Sarabhai, were as much motivated by
political desires to create a unified community of Indian television as by
the economic demands of mobilizing new technologies like satellite communications to represent and realize these goals.[47] Ashish Rajadhyaksha
argues that Sarabhai was convinced of the technological potential of satellite communications to eradicate the problems of illiteracy by providing
educational materials in an "audio-visual" format, and to overcome the
hurdles of geographic distance and linguistic diversity "simply with its 'cred-
ibility' and 'rare persuasiveness.'"[48]

Sarabhai's vision of using television for social development provides
subtle insights into some of the assumptions that the nationalist elites in the
1960s made about the power of satellite communications. He persuasively
argued for an Indian space policy that would be consistent with the Gandhian ideals of self-reliance in the domestic context and the Nehruvian ideals
of nonalignment in the international arena. Sarabhai's contributions toward
the development of the ISRO were indeed impressive. Like Nehru, Sarabhai
was a staunch modernist who also was drawn to Gandhian ideals of *swaraj*
and *swadeshi*. Following the Nehruvian model of synthesizing a mixed vision
of nationalism, Sarabhai did not hesitate to collaborate with both the Soviet Union and the United States in his efforts to design an Indian National
Satellite (INSAT) system.

When Sarabhai died in December 1971, many in India were concerned that the government of India would not develop the INSAT system further. However, on September 18, 1969, when Sarabhai was still at the helm of ISRO, the government of India had signed an agreement with the United States government to study the role of satellite television for national development.[49] Called the Satellite Instructional Television Experiment (SITE), the project gave the scientists at ISRO an invaluable opportunity to use NASA's Applications Technology Satellite, *ATS-6*, for a period of one year. Although Sarabhai's plans for INSAT took a back seat to the SITE experiment, the use of satellite communications for national development remained a priority for the scientists at ISRO during the 1970s.

While working with NASA on the SITE project, India launched its first research satellite, *Aryabhata*, using a Soviet rocket, in 1975. At the same time, ISRO signed an agreement with the Franco-German Symphonie satellite administration to conduct the Symphonie Telecommunications Experimental Project (STEP) from 1977 to 1979. The result of these international collaborations was that India had developed the most sophisticated space program in Asia by the end of the 1970s.

However, Brian Shoesmith finds that the impact of these early efforts dissipated slowly because ISRO's research was very much oriented toward national development projects and was heavily bureaucratized, due to the centralized authority exerted by the officials of the government of India.[50] Not unlike many other critics of India's early efforts into space research, Shoesmith argues that the SITE experiment "exemplifies the situation that developed."[51] The SITE project was conducted in 2,400 villages in five different states across the country in 1975–76. The proclaimed goal of the SITE project was "to gain experience in development, testing and management of a satellite-based instructional television system, particularly in rural areas, to demonstrate the potential of satellite [technology] in developing countries, and to stimulate national development in India . . . to contribute to health, hygiene, and family planning, national integration, to improve agricultural practices, to contribute to general school and adult education, and improve occupational skills."[52]

The SITE project was instrumental in setting the political, economic, and cultural agenda for the use of satellite communications for national development in postcolonial India. For instance, in a study of international funding agencies in the United States, entitled *Technical Assistance in Public Administration Overseas*, Edward W. Weidner identifies India as one of the few Asian countries where the government has the necessary resources to implement development projects in their national communities.[53] Similarly, in a report prepared for the Agency for International Development, Wilbur Schramm and Lyle Nelson present India as an ideal case study for the use of communication satellites for education and development.[54]

However, a dominant theme that constantly emerges in these discussions about the use of satellite television for national development in India is that the process of communication is seen as a simple linear process of transmitting information from the "experts" to the uninitiated "masses," who, it is assumed, strive to be more "productive" like the former. The emphasis continues to be on the modern experts' ability to transfer appropriate scientific and technical knowledge to the laity by training them to identify the ills of underdevelopment as a set of variables that can then be corrected using "scientific" techniques of social engineering. Moreover, as Luis Beltran reminds us, when these social-scientific models are extended to the international level, it is assumed that the Western nations of Europe and North America, which pioneered the expertise and know-how in the use of satellite communications, are the ideal types to be emulated by the peoples of the Third World who are engaged in the work of national development.[55]

Although the development of the INSAT system depended heavily on the resources and expertise of scientists from NASA, Western Europe, and the Soviet Union, the government of India recognized, as Raman Srinivasan puts it, that there is no such thing as a "free launch."[56] On April 10, 1982, India launched its first multipurpose communication and meteorology satellite, *INSAT-1A,* from Cape Canaveral in the United States. Although the satellite worked only for a few months and had to be abandoned due to technical problems in September 1982, *INSAT-1A,* as the name suggests, was the first in a series of communication satellites that India had planned to launch in the years ahead. On August 30, 1983, *INSAT-1B* was launched, once again from Cape Canaveral. Well into the 1990s, *INSAT-1B* served as India's premier communication and meteorological satellite. It was removed from orbit in August 1993, by which time a new INSAT-2 series of satellites had been introduced into service. With the launch of the INSAT-3 series on March 22, 2000, Sarabhai's ambitious vision for creating national satellite system with a distinctly "Indian" character was now heading into the twenty-first century.

The guiding framework for the nationalist desire to create an "Indian" alternative to counter the dominance of Western satellite systems is encapsulated in the historic debate about the need for a New World Information and Communication Order (NWICO) during the 1970s. Initiated by the

Non-Aligned Movement and sponsored by UNESCO, the NWICO debate was a valiant attempt to redress the imbalances in international communications. During the heyday of the Non-Aligned Movement in the 1970s and 1980s, many Third World nations emphasized the need for an alternative framework to create a more equitable exchange of technologies, expertise, and resources in a world dominated by their erstwhile colonizers. The creation of the Non-Aligned News Pool in 1976 represented a significant step toward redressing imbalances and biases in news flows around the world. D. R. Mankekar, who served as the first chairman for the News Pool, argued that the four big wire services in the West—Associated Press, United Press International, Reuters, and Agence France Press—had a monopoly over news reporting in the world. India, he argued, had taken the lead in the creation of the Non-Aligned News Pool so that Third World nations could develop and share their indigenous news media sources as viable alternatives to the dominance of the Western agencies in the world.[57]

In 1980, an international commission established under the leadership of Chairman McBride published its report, entitled *Many Voices, One World,* which argued in favor of creating a more egalitarian and equitable order for information and communications flows around the world.[58] Such was the power and sway of the NWICO debate in the 1970s that leading proponents of the Western model of international development proclaimed its death and posited the rise of a "new paradigm" that, they argued, took into account the realities of inequities in power relations among the First, Second, and Third World nations.[59]

Although the celebration of a New World Information and Communication Order and the proclamation of a "new paradigm" in its wake did address the problems of inequities and power relations, the attempts to redress their daunting political realities left much to be desired. For instance, the First World nations, heartily embracing the notion of "free flow" of information, resisted any kinds of curbs on the freedom to exchange information around the world. On their part, the nonaligned nations, often with strategic support from the Soviet Union, were inclined to accept the freedom to exchange information globally, only if this exchange was a "fair" one. Given the realities of power inequities among nations, they remained committed to take all measures to protect their national security and cultural sovereignty in the New World Information and Communication Order.[60]

Faced with irreconcilable differences in the definitions of "freedom" and "fairness" in the international arena, the Non-Aligned Movement's revolutionary initiative for free and fair flow of information in the "new world order" was doomed to failure. The reasons for this failure are many. For one, the initial euphoria of postcolonial independence was already a distant memory for many African and Asian countries. Given the political and economic compulsions of national development, many Third World nations were consumed by an overwhelming desire to "catch up" with the West in the international capitalist order. Many African nations fell to dictatorships and military regimes. Led by the United States, Western nations often imposed enormous political-economic pressure on many radical regimes. Slowly but steadily, the "mixed visions" of national development in the postcolonial world were consumed by neocolonialist visions of capitalist development. As the idea of a single world capitalist system has become "enormously powerful and truly global," it has "opened fresh territories for the spread of capital."[61] Therefore, Prakash argues,

> Massive movements of capital and migrants have turned some areas into "emerging markets" while marginalizing others into "basket cases." The structure of flexible sourcing and markets has scrambled older divisions, producing "Third World" enclaves in Los Angeles and New York, generating "First World" capitalist "miracles" in East Asia. As the internationalization of capital after the collapse of the Soviet Union produces a new, post-nation-state organization of class structures, advancing the process of dismantling the Fordist combination of big capital/big labor/big government in favor of flexible accumulation, it also seeks to turn international organizations into capital-servicing units.[62]

Following the collapse of Soviet communism, and after the end of the Cold War, the integrated world of global capitalism formed by a peculiar coalition of Western capitalism and the "mixed economies" of Third World nationalism has only become more peculiar because of its affiliation with the emerging markets of the postcommunist world. Signaling this change in the most emphatic terms, Intersputnik, which was launched by the Soviet Union as a noncapitalist organization for international space communications in 1971, began commercial operations of its satellite systems in 1992. While the website of Intersputnik now looks similar to that of any Western transnational corporation, the company claims to "offers its clients state-of-the art satellites with high-power transponders, providing customers with outstanding performance and reliability."[63] Since 1992, several Indian television networks, such as Asianet, Sun TV, and Zee TV, have leased Intersputnik's transponders on the Gorizont series satellites to beam their programming across the subcontinent. Although many of these networks have now moved to other commercial systems, like AsiaSat and PanamSat 4, Intersputnik is heavily investing its resources to expand its satellite footprints in the lucrative Indian television market. On March 9, 2000, the Indian government announced its decision to become a member of the Intersputnik system so as to enable its national telecommunications authority, Videsh Sanchar Nigam (VSNL), to participate in the international organization's commercial operations

Even Western corporations like Lockheed Martin, which were not allowed to collaborate with Intersputnik in the Cold War era, are now eager to sign contracts to manufacture and operate communication satellites for the formerly communist organization. Intersputnik's new generation LM-1 satellites, built by Lockheed Martin, now orbit the earth to provide coverage to any television network that seeks to broadcast over an area from the Middle East to Southeast Asia.[64]

In this topsy-turvy world of satellite communications, what do we make of the paradigmatic "communications model" proposed by Lerner for the modernization of traditional societies in the postcolonial world? As triumphant moderns of the First, Second, and Third worlds appear united in their collaborative ventures, the ambivalent transitionals have gained a new-found legitimacy as potential consumers in a new television market. The ideological ambivalence of the transitional audiences toward the international capitalist order—which was once deemed potentially illegitimate and a matter of grave ideological concern during the Cold War—now represents a challenge with enormous profit potential for any television network willing to risk its economic resources. As the "moderns" and the "transitionals" join hands to develop more efficient satellite communication systems to deliver newer, hybrid programming formats around the world, the "traditionals" appear to be the only postcolonial audiences who occupy the illegitimate category of "underdevelopment" in the project of international development that began in the heyday of the Cold War era.

However, in the post–Cold War world, the category of "traditionals" no longer seems to be threatened by the ideological dangers of global fascism or international communism. Instead, it is seen as a category of underdeveloped audiences who are drawn to regressive and antiquated traditions in their own cultures that are defined by the moderns of the world as "fundamentalism," "tribalism," and "terrorism." In this global landscape of electronic capitalism, Homi Bhabha argues, "nationalist awareness and authority has been brutally asserted on the principle of the dispensable and displaceable presence of 'others' who are either perceived as being premodern and therefore unde-

serving of nationhood, or basically labeled 'terroristic' and therefore deemed
unworthy of a national home, enemies of the very idea of a [*sic*] national
peoples."[65]

While recognizing that the "unprecedented advance in the internationalization of capital" is a matter of great concern to those who seek to protect
the rights of subaltern groups and marginalized communities, Prakash advocates caution and skepticism toward "apocalyptic visions announcing the
end of the world" in the "last stage of capitalism," its "final general crisis"
and "its spread to every corner of the globe."[66] These announcements, Prakash
argues, "either diffuse criticism or postpone it to the time of the future ca-
tastrophe" and thus ignore the hybrid life of capital itself.[67] To acknowledge
hybridity in this context is not to speak of ideological resistance to the international spread of capitalism or to embrace theoretical *jouissance* (enjoyment)
that can overcome the homogenization of global cultures. Rather, the recognition of hybridization is an acknowledgment of the differentiated and dif-
ferentiating structure of electronic capitalism: it "operates in unevenness, and
it proceeds by domesticating difference."[68]

In postcolonial contexts, such as India, where developmental notions of
Western capitalism, Soviet communism, and Third World nationalism have
been "squeezed dry of their emancipatory potential," Prakash argues that
there is an urgent need "to think through and beyond established forms of
politics and knowledge . . . along differentiated, interpellated, mobile and
unsettling lines."[69] The lines along which he invites us to intervene in the
politics of hybridity are "those relocations of dominant discourse" that emerge
"not from the nation-state, not from the Third World space, but from contingent, contentious and heterogeneous subaltern positions."[70] As I will show
in the next two chapters, the rapid growth of electronic capitalism in India
has unsettled many of the legacies of Western colonialism and postcolonial
nationalism that have long enforced an artificial choice between tradition and
modernity.

4   "Gandhi Meet Pepsi": Nationalism and Electronic Capitalism in Indian Television

"GANDHI MEET PEPSI," declared the headline for an article written by the noted feminist scholar Urvashi Butalia (1994) in the *Independent on Sunday*.[1] The headline is heady, the contrast clever, and the significance stunning. The superstar of Indian nationalism forced to face the rising star of transnational consumerism; the authorial symbol of the struggle for independence of the great-grandparents' generation asked to accede to the chilling choice of a new generation's transnational interdependence; the deified signifier of Orientalist renunciation compelled to make room for the reified significance of Occidentalist consumption. One could go on, but the ingenious headline makes the conclusion crystal clear even before one pores over the fine print. But for the ingenuous reader, the lead line says it all: "Western culture is sweeping India," Butalia announces, with exaggerated effect: "India, which contains one-sixth of the world's population, is no longer aloof and mysterious. It has dismantled its trade barriers and liberalised its economy. The philosophies of its modern founders, Gandhi and Nehru, seem as far away as Sanskrit texts. We have satellite television. We are a market. We consume."[2]

"Gandhi Meet Pepsi." The equation seems strikingly simple. To liberalize is to be liberated. To be liberated is to undergo revolution. To undergo revolution is to follow the lead of a new star. India has a new star in satellite television. It is the choice of a new generation—Generation NeXt. Never mind the reification—India has been plugged into the transnational networks of electronic capitalism that have consumed large parts of Asia and much of the world since the 1990s. In this chapter, I address the question of how an Indian *community* of television is imagined by overwriting the narratives of nationalism in the discourse of electronic capitalism. I define *overwriting* as a paradoxical dual process of writing and erasure. By deconstructing the simultaneous writing and erasure of the subject of nationalism in electronic capitalism, I address how Gandhi's name is symbolically overwritten to death by competing visions of transnational corporations like Pepsi in postcolonial India.[3]

Waiting to Consume

"Gandhi Meet Pepsi." The rhetoric is rhapsodic, the climate chaotic. The scenes are rather familiar; only contexts radically differ, as the magic mantra of "globalization" is chanted by *New York Times* columnists writing about the changes they see in cities and villages across Asia. As the *New York Times* Asia correspondent Philip Shenon discovers, the transnationalization of national communities across Asia has engendered a "race to satisfy TV appetites."[4] In Asian countries, poor and rich alike, Shenon sees "parabolic dishes" as often as "rice paddies," and he captures the following "scenes from the satellite television revolution," in its early days yet:

> In Hanoi, officials of the Vietnamese Communist Party and foreign guests in some of the city's larger hotels watch the BBC's hourly news, beamed live by satellite from London. In Hong Kong, stock traders end their day in the office by flipping on Cable News Network to catch the opening share prices half a world away in Europe, while at home their children sit glued to a Madonna video festival on MTV. In Indonesia, rice farmers on the island of Sumatra gather around a communal television set to watch the American soap opera "The Bold and the Beautiful"—they don't understand the English script, but they seem to enjoy it anyway. . . .
> In the prosperous city-state of Singapore and north across the Gulf of Thailand to Bangkok, viewers pay a monthly fee to enjoy recent Hollywood films on a new pay television service, Home Box Office Asia.[5]

Over in China, Nicholas Kristof argues that satellite television has brought an "Information Revolution" to the country, and he paints the following picture for the readers of the *New York Times:*

> hundreds of thousands of satellite dishes . . . are sprouting, as the Chinese say, like bamboo shoots after a spring rain. Already millions of Chinese can hook in via satellite to the "global village," bypassing the Communist Party commissars and leaving them feuding over how to respond."[6]

Further across in India, the *New York Times* correspondent Edward Gargan finds that "TV Comes in on a Dish," and Indian viewers are gobbling it up:

> For decades, Indians have been restricted to the fare served by state television, known as Doordarshan, a diet of tedious discussions by Government bureaucrats, ancient Hindi movies, generous amounts of singing and dancing, and news programs often distinguished by their heavy censorship and age. But . . . [in 1991] new channels began appearing on Indian televisions—MTV rock videos, an all-day sports channel and . . . the BBC's new World Service Television . . . courtesy [of] . . . STAR TV. For India, a nation long padlocked to the Government's version of reality, the candy-store variety of programming has brought a poorly contained giddiness.[7]

Even in the blurry snapshots of "scenes from the satellite television revolution in Asia" that the *New York Times* correspondents manage to capture, an interesting narrative begins to emerge where nationalism and electronic capitalism insatiably feed into each other: a race is on to "satisfy the TV appetites in Asia" where "parabolic dishes" are "as common a sight as rice paddies," "sprouting . . . like bamboo shoots after a spring rain." So when "TV comes in on a dish . . . India gobbles it up." And why not? The restricted "fare served by state television" has been a "diet . . . [so] tedious" that the government's version of reality has become blurred. Now, "courtesy [of] . . . STAR TV," millions in India are suddenly witness to the "candy-store" vision of consumption that transnational television network brings, and with it comes a "poorly contained giddiness."

Sigmund Freud would perhaps have diagnosed the "poorly contained giddiness" of the hungry Indian peeking at the candy-store vision of transnational television as the outcome of an alienated ego struggle between base feelings of the id and elevated ideals of the superego. Karl Marx would have criticized it as the false consciousness of a societal base being force-fed the ideology of global capitalism. They might well be right. Yet, after countless sessions in the name of psychoanalysis and numerous resolutions in the name of Marxism the world over, the hungry rarely seem to follow Freud and Marx.

Perhaps the hungry follow Friedrich Nietzsche, who would enjoin them to ask with gay abandon, "Why attribute the giddiness at the candy store to an invisible unconscious or a false consciousness, when the world appears ours to consume?" To some journalists like John Rettie of the *Guardian*, it appears that the "giddiness" induced by transnational television can be overcome by uninhibited partaking, only in the best of spirits.[8] So linger, if you will, after a weekend dinner at the posh Maurya Sheraton in west Delhi. The awaiting vision will shock you so that "you may need the fortifying power of the hotel's wildly expensive imported cognac," writes Rettie.

> For around midnight, an incoming tide of beautiful women wearing skin-tight dresses, plunging necklines and micro-skirts sweeps into the lobby. If and when you recover your breath, you may well ask your host: "How can a five-star Indian hotel let these women ply their trade so blatantly?" "No, no, you don't understand," he will laugh, pointing out the young men in the women's wake. "These are the children of the Indian elite, and they're just coming to the Saturday night disco."[9]

If the "wildly expensive imported cognac" works its magic, and you sufficiently recover from the shock of what you see, you will realize that these youngsters "and their families form only a minuscule layer at the pinnacle of society."[10] But you will also recognize that India's sizeable middle class—estimated to be anywhere from 100 to 250 million people—who see this giddy vision of transnational consumption on television every day and night "want to emulate" this lifestyle.[11] Whatever the numbers are of the Indian middle class that is waiting to consume, "television hammers home the message: consumption is an inalienable right. Economic reforms and market liberalisation are the new religion. The prime minister, P. V. Narasimha Rao, and the finance minister, Manmohan Singh, are its gurus."[12]

In the general elections of 1991, when P. V. Narasimha Rao's Congress Party came to power, India was in the midst of a severe economic crisis. The Soviet Union, the second largest export market for India, and an invaluable trading partner that provided cheap military hardware, scientific technology, and petroleum products, was falling apart. Iraq's sudden invasion of Kuwait put yet another petroleum-rich market in turmoil, and India was forced to expend exorbitant amounts of its fast-dwindling foreign exchange reserves to buy much-needed oil. With Iraq's occupation of Kuwait, India suddenly lost about $2 billion in annual remittances from Indians working in the Gulf states, as they fled for home in fear. With one economic crisis mounting over another, India nearly ran out of money to pay for even its most essential imports like oil and was dangerously close to defaulting on loan payments to international financial institutions like the World Bank and the International Monetary Fund (IMF). The economic crisis was temporarily averted by pledging India's gold reserves in the Bank of England in exchange for hard currency.[13]

However, the symbolic significance of the nation pledging its gold reserves for cash was hardly lost in a country where pledging familial gold is traditionally seen as the ultimate acknowledgment of one's economic humiliation. However, as the national economy gained temporary respite, many stalwart politicians—from right-wing conservatives to left-wing communists—momentarily subsumed ideological differences and applauded Prime Minister Rao for his deft handling of a difficult situation. They commended Rao for boldly breaking with tradition by offering the key post of finance minister to Manmohan Singh, a career bureaucrat and a nonpolitician.

Singh, a Cambridge-educated economist and a former governor of India's central bank, was seen as a man with a mission: to dismantle the nation's infamously corrupt "License Raj" economy, which had too many governmental, bureaucratic, and political hurdles for foreign investors, and to make Indian markets lucrative for global investors and corporations. His plan to change the economic equation was daring in its striking simplicity: let global investors invest however and wherever they want to. His promise to the transnational corporations was clear and plain: no political strings attached; no bureaucratic red tape to wade through.

The significance of the changing equation in India was not lost on transnational corporations like Pepsi. In fact, early in 1989, two years before Rao's economic reforms eased the way for global competitors, Pepsi and its archrival Coca-Cola were among the first few transnational corporations eager to set up shop in the lucrative Indian markets. Coca-Cola's proposal to set up an export-oriented project in the Noida export-processing zone near the nation's capital, Delhi, was rejected by the government at the time, and Coca-Cola was not able to set up shop in India until four years later, after the consolidation of the process of economic liberalization.[14]

On the other hand, Pepsi's proposal establishing food-processing and soft-drink-production facilities in India was approved in 1989 and was even increased in scale to include, among other things, the processing of 90,000 tons of tomatoes for canning and export.[15] The government approved Pepsi's entry into India on the condition that the transnational corporation would export a part of its production, create local jobs, and upgrade existing technology in the high-priority food-processing industry. Pepsi's promise of increasing exports from India and bringing in some much-needed foreign exchange had to be rather alluring to a government tottering at the brink of an economic crisis in 1989.

The potential for increasing local jobs is always very appealing—especially so in an election year—in a country like India, which has been struggling with a staggering unemployment rate for decades. The promise of bringing in the latest technological know-how to the food-processing industry is very seductive in a nation that is prone to climatic vagaries and has few storage facilities or techniques to save the excess yields of a good monsoon year for the following dry season. In this context, the "food-processing revolution" that Pepsi promised in the 1990s appeared to be an obvious solution to the peculiar poverty India suffers, even as it enjoys agricultural riches, thanks to the Green Revolution of the 1970s, and a dairy surplus, thanks to the White Revolution of the 1980s.[16]

In spite of—or perhaps because of—the fact that Pepsi's quiet arrival in India predates the celebrated economic liberalization of the 1990s, it has not been without its share of controversies. While there was little criticism of Pepsi's food-processing projects, its soft-drink plants quickly became a subject of intense political debate. Even as Pepsi was beginning to negotiate the terms of access to the Indian markets in 1989, it was involved in a marketing and distribution controversy with a domestic bottling company, Double Cola. Although embroiled in controversies from the beginning, Pepsi lost little time in establishing its world-famous brand name as "the choice of a new generation" in the vast Indian markets.

In positioning itself as a "generation neXt" product, Pepsi tried to attract a younger, postcolonial generation in India who might find little to identify with in the nationalistic obligations of an earlier generation, and much attraction to the transnational brand of uninhibited consumption prominently advertised on foreign and domestic television networks. Using any and all locations, from billboards to bus panels, trash cans to television commercials, Pepsi periodically launched a series of advertisement campaigns to appeal to the consuming potential of a new generation of youngsters. While the economic significance of aligning with the young generation in India is considerable for Pepsi, the symbolic significance is staggering.

As Neel Chatterjee, the general manager of Pepsi Foods, admits, the economic gains to be made are significant because the young have "tremendous purchasing power." Warming up to Pepsi's theme of being "the choice of a new generation" and its symbolic significance in relation to the Indian youth, Chatterjee argues: "The teen-agers today are more perceptive, know more about the world, know what is good for them and what is not. The awareness among teenagers is growing unlike the past, when it was learning lessons with some occasional outings or a game. Today they have mustered the courage to stand up for what is needed and thunder that they have the right to choose."[17]

The marketing hype aside, there is something to what Chatterjee says about the changing character of Indian youth, particularly among the more modernized, urbanized, and Westernized segments of the population. Walk into a swanky discotheque in Mumbai, as Clarence Fernandez of Reuters did, and you will find the new generation of "young achievers of India's urban middle-class" dancing to the latest tunes, brought to you, of course, by Pepsi.[18] You are perhaps in the midst of one of the many nightlong "raves" sponsored by Pepsi in major cities in India. As you approach the gyrating figures beneath the flashy strobe lights, you may find someone like Uday Sharma, an advertising executive, who tells Fernandez that the Pepsi "raves" are "a good way to relax and cool off after a hard day's work."[19] For an evening's enjoyment, Sharma thinks nothing of paying an entry fee of 750 rupees (approx. $24). In India, 750 rupees is still a princely sum for many "like the hand-cart pullers at the city's Crawford Market, who earn only a few rupees a day and sleep in rows on the pavements every night."[20] Thus, Fernandez finds that in India, signs of "the fast-accelerating wave of consumerism" like the Pepsi rave he is witness to "contrast sharply with the more ascetic times of independence leader Mahatma Gandhi."[21]

"Gandhi Meet Pepsi." The equation is no longer simple. A naked truth of Pepsi's transnational brand of consumption is fleetingly revealed, as a stark contrast between the insatiable hunger of the "haves" and the hungry "have-nots" is brought into relief by the visions of excess that economic liberalization has heralded in postcolonial India. Less than half a decade after the inauguration of economic liberalization, one could witness a marked shift in popular perceptions about the role of transnational corporations in the national community. Advocates of the Pepsi brand of transnationalism have argued that while the economic liberalization initiated by Rao and Singh in 1991 was a good start, political pandering to socialist populism, rampant corruption, and bureaucratic red-tape-ism continued to frustrate people's ambitions. Reflecting this perception in a survey of the Indian economy, Emma Duncan, South Asia editor of the *Economist,* reports on the results of economic liberalization in India thus:

> India has an economy slightly smaller than Belgium's. Its GDP per head is $310. Fewer than half of its 950 m[illion] people can read. Between them, they have only just over 6 m[illion] telephones and 35 m[illion] television sets. Some 14 percent of the population has access to clean sanitation—a lower proportion than anywhere else except for a handful of Sudans and Burkina Fasos. According to the World Bank, 63 percent of India's under-five-year-olds are malnourished. Perhaps 40 percent of the world's desper-

ately poor live in India. Apart from a tiny elite in Delhi and Mumbai, India's rich are poor by anybody else's standards: according to the National Council for Applied Economic Research, only 2.3 percent of the population has a household income of more than 78,000 rupees ($2,484).[22]

Statistics, one may argue, are damned lies. But even damned lies can tell an appalling tale well. As the *Economist* survey tells us, there is little to celebrate for many in India who view economic liberalization as a distant vision on their television screens, and for many others who only get to see an occasional flicker in the local stores in their communities and the newspaper and magazine stalls in their neighborhoods. Therefore, Duncan argues, some of the recent changes in India are, in fact, the result of another recent revolution, "about which the government could do nothing but fume: the advent of proper television."[23]

With the media revolution that "proper" television brings to India has come "an enlivening consciousness" among people of how "badly the country has done" and how much it needs to do "to catch up with the rest of the world." Thus, in offices and companies across India, "standard cramped, grubby office[s]" are giving way to "something more in tune with international taste." "Horribly-cut casual clothes" are being cast aside "in favour of suits and ties." Outmoded cars are being replaced by "flashier" ones, and parties are becoming "more ostentatious."[24]

Duncan finds that television gets the blame and the credit for many of the changes taking place in Indian consumer culture. Many girls are now discarding the traditional Indian *sari* and *salwar kameez* in favor of Western clothes. Some "advanced cosmopolitan couples" are now "living together before they are married" because, Duncan argues, "it's easier to explain it to a mother who has seen *The Bold and The Beautiful* on STAR TV."[25] Breasts are starting to appear prominently on covers of glossy magazines, and many Indians are beginning to get comfortable with the excessive materialism of consumer culture. Yet, Duncan reminds us, "consumerism" remains "a troublesome idea" in India, whose national hero "liberated his country wearing a loin-cloth, and spent his leisure hours toiling at a spinning-wheel."[26]

"Gandhi Meet Pepsi." The equation becomes troublesome as the ideological collisions and collusion between nationalism and electronic capitalism are dramatically displayed on television screens across India. A stark contrast between the hungry "haves" and the hunger of the "have-nots" that is brought into relief by the specters of Gandhian nationalism must now be acknowledged by the nationalist elites in order to sustain the legitimacy of the Pepsi brand of consumption in India. In an attempt to rewrite the troublesome equation, the former commerce minister (and later finance minister) P. Chidambaram argues: "There's no question of haves and have-nots" in India today.[27]

Chidambaram, one of the most articulate proponents of the Indian government's policies of economic liberalization, is too well versed in the variety and diversity of class, caste, and other social hierarchies to ignore their many contradictory trajectories. After all, a vast majority of India's population, now past the one-billion mark, can barely afford a peek at the "Pepsi" vision on television, let alone reap the riches that electronic capitalism promises to deliver into households across the national community. Yet, even as he acknowledges this daunting reality of economic inequalities, Chidambaram sees no reason to curtail the growing power of transnational corporations in India's new consumer culture. His argument goes thus: "We're talking of a market of about 300 million people. That's larger than many countries. Just because most people can't afford them is no reason to deprive everybody of good-quality stuff."[28]

Others take a different approach in explaining the troublesome equation. Mani Shankar Aiyar, among the more vocal supporters of economic liberalization in the Congress Party, seeks to relegate the confusing equation to the realm of myth when he says: "The foreign invasion of our consumer goods market is a myth. Foreign investment approvals in mass consumption items and services such as Kentucky Fried Chicken and Pizza Hut is less than two per cent of all approvals. . . . Much less than one-tenth of approved investment in industry and infrastructure is from outside the country, and actual investment has been less than one-twentieth."[29]

Both Chidambaram and Aiyar may well be right in their well-articulated rationales for supporting the politics and policies of economic liberalization, given the overwhelming momentum of electronic capitalism in Indian consumer culture since 1991. In terms of a statistically consistent analysis, Aiyar's argument is sound. Even a cursory look at the breakup of foreign collaborations and investments in different sectors of Indian industry from August 1991 to April 1995 reveals that out of a total number of 6,196 approved investments, only 362 are in the food-processing sector, thus constituting a mere 6.2 percent. Even in terms of total foreign investments in all industrial sectors in India, statistics reveal that in sheer numerical terms, actual foreign investments vis-à-vis approvals are only a trickle, rather than the deluge that the fashionable trends of foreign products seen on television suggest.[30]

The reasons are, of course, many. India is only one among the many emerging markets available to media networks and transnational corporations seeking to establish their businesses around the world. India's immense consumption potential notwithstanding, many media networks, like STAR TV, and transnational corporations, like Pepsi, have looked to invest in the more established systems of China and Hong Kong, the emergent economies of Southeast Asia, and the strategically significant markets of Russia and its former satellite communist nations. Sound economic logic suggests that for the liberalization policy to succeed, India must become more competitive with other emerging markets to invite large-scale investments by media networks and transnational corporations.

In this context, as Chidambaram recognizes—in a large, diverse, democracy like India that has embarked on a policy of economic liberalization—it becomes a political suicide, if not a practical impossibility, for any government to prohibit 300 million thirsty people from partaking in the Pepsi culture just because an equally large number of people may be thirsting for pure drinking water, or a comparably large number may be struggling to keep their families from going to bed hungry every night.

Moreover, in a representative democracy, the sound logic of economics must come face to face with the pressing demands of politics. There are times when neatly designed statistical models, nicely planned economic ideals, and clearly articulated political rationales turn topsy-turvy and transform into ugly displays of power and violent conflicts of ideologies. In India, such times come frequently; much too frequently, some would argue. But, most significant, they come at least once every four or five years, as political parties prepare, rather elaborately, to campaign for the general elections to the Lok Sabha, the lower house of the bicameral Indian Parliament. In such times, the economic potential of the consuming desires of a powerful third of India's one-billion-strong population must submit to the political significance of the pressing needs of the remaining two-thirds. In the summer of 1995, a politician somewhere acknowledged the significance of this harsh reality of electoral politics with blunt honesty in the runup to the general elections of 1996: "they may not drink Pepsi, but they do vote."

When the Nation Writes Back

Like many previous general elections in India, the crux of the runup to the 1996 elections was the track record of the powerful Congress Party. By the summer of 1995, as political parties began charting out their strategies in earnest, it became apparent that the bone of electoral contention would be the economic liberalization policy, which appeared to have lost some of its initial fizz. To revitalize the fading memory of the "achievements" of economic liberalization, the Congress Party launched a media blitz on the state-sponsored network, Doordarshan, which it controlled by virtue of being the party in power in the central government. The Congress Party was not shy about using the state-owned electronic media to inform the public about the accomplishments of its policies of liberalization, which, it claimed, had rescued the nation from the verge of economic collapse. The government's media campaign included

> propaganda films such as the one that shows a villager queuing up at a public-distribution outlet from where everyone returns satisfied. Or a 30-second clip where a city bound villager is persuaded by a soothsayer that a bright future awaits him in the village itself—courtesy Jawahar Rozgar Yojna (JYR) [the Congress Party's ambitious rural employment program].... A bank of ... docu-dramas and clips—some commissioned [by Doordarshan] and others by various ministries such as the Rural Development Ministry (RDM) and the Health Ministry.[31]

Despite the intense media campaign by the Congress Party, it was clear that the effusive support that Prime Minister Rao's innovative policies had received after the general elections of 1991 appeared to be a fading memory—even among the supporters of economic liberalization. Pollsters and political pundits who could sense a general discontent among the electorate across the country perceived it as a critique of the failings of the economic liberalization policy launched by the Congress Party in 1991. In nationwide opinion polls, the electoral pulse seemed apparent. The rising ranks of urban middle classes, who generally support the government's policies of economic liberalization, seemed unhappy with the slowing pace of reforms. Many appeared to find fault with the Congress Party for falling prey to what they considered to be the regressive politics of government subsidies for the agricultural rich and naive socialist policies of state-sponsored development to uplift the rural and urban poor.

Furthermore, influential leaders of the business community found fault with the government for its inability to stem political corruption and bureaucratic rot in the Indian economy. This unhappiness was situated as a larger critique of the Congress Party for not living up to its promise of economic liberalization and for surrendering to the political compulsions of competing with the ideologies of xenophobic nationalism and egalitarian socialism being preached with remarkable success by the radical fringes of both the right-wing and left-wing political parties.

Among the remaining two-thirds of the national community—who share little in common with the top one-third—are many voices that are rarely surveyed by pollsters. In the rare event that these silenced sections of the society—the subalterns, if you will—found stray voice in the plethora of polls conducted by the national media, there was a perceptible sense that economic liberalization is a "prorich" (and therefore antipoor) policy. The poor and the propoor sections of the society appeared to perceive liberalization as a policy that panders to the desires of the powerful middle class, promotes the interests of the rich and corrupt, and fails to serve the swelling numbers of needy and poor. This perception was interpreted by pollsters and political pundits as a larger criticism of the Congress Party for surrendering the nation to the Pepsi brand of transnational consumption, and for straying from its avowed principles of Nehruvian socialism, for example, equitable development, and the cherished ideals of Gandhian nationalism, like *swaraj* and *swadeshi*.[32]

"Gandhi Meet Pepsi." The equation becomes increasingly complex. The competing visions of postcolonial nationalism and electronic capitalism seem increasingly in conflict, as the fundamental ideological contradictions of politics and economics surfaced in the demands of electoral necessity in the general elections of 1996. To win the elections, the Congress Party was induced to promote economic liberalization as a transnational dimension of its Gandhian and Nehruvian nationalist ideals. However, this strategy runs the political risk of alienating the vast majority of the electorate that is disillusioned with the excesses of transnationalism and finds compelling reasons to turn toward the xenophobic nationalism being preached by radicals in right- and left-wing parties.

To counter the radical rhetoric both from the right and left fringes of the political spectrum, the Congress Party was forced to contradict its own policies of economic liberalization and to proclaim its nationalist legacy of Nehruvian socialism and Gandhian ideals of homespun nationalism. Yet the contradictory assertions of policy pronouncements and political ideals by the Congress Party did little to enhance the confidence of powerful segments of the middle class—business and industry leaders who support greater liberalization of the Indian economy. In the ensuing media spectacle of its contradictory electoral campaign, the Congress Party found itself run over by a juggernaut of its own making. It was forced to confront the contradictions of its transnationalist ambitions and nationalist compulsions, even as it had to proclaim its legacy of Gandhian-Nehruvian principles, which had enabled it to dominate the Indian electorate like a colossus for over forty years.

Opponents of the Congress Party's economic liberalization policy were acutely aware of its electoral quandary, as they charted out their own media campaigns in the runup to the elections of 1996. Many of the media campaigns launched by political parties opposed to the Congress Party were aimed against the Pepsi brand of transnational consumption that was evident in the policies of economic liberalization. Among the more visible of these media campaigns was one led by the firebrand socialist leader George Fernandes, who, as a government minister in 1977, ousted Coca-Cola for not revealing the "secret formula" to domestic manufacturers in its Indian production plants. At a meeting held in Delhi in July 1995 to mark 500 days of campaigning against transnational corporations like Pepsi and Coca-Cola, supporters of Fernandes lauded the efforts of protestors, who, for 500 days, had blackened Pepsi and Coke billboards across the country. "The . . . (campaign) for the ouster of Pepsi and Coca Cola was launched as a symbolic protest against the entry of TNCs [transnational corporations]. Their ouster from India will give a clear signal to other exploiting TNC's to pack up or face the united might of the Indian people," a declaration issued at the end of the meeting proclaimed.[33]

In May 1995, two months before the 500–day anniversary meeting, Fernandes successfully raised concerns against TNCs in Parliament. Several parliamentarians from socialist parties, like the Janata Dal, communist parties, and regional parties, and even some members of the ruling Congress Party supported Fernandes as he blasted the twin symbols of transnational consumerism: Pepsi and McDonald's. What provoked Fernandes's stringent attack this time was the government's approval for the launch of fast-food outlets by Coca-Cola's fast-food partner, McDonald's, and the Pepsi subsidiary Kentucky Fried Chicken (KFC). According to this plan, approved by the government of India, the first of thirty KFC outlets was to be opened in the southern city of Bangalore in June 1995, while McDonald's planned to launch its sixty outlets in the following year. With the government approvals for Kentucky Fried Chicken and Big Macs coming on the eve of a crucial general election, several parliamentarians began to smell electoral trouble in the fast-food chains. In a memorandum to Prime Minister Rao, fifty parliamentarians across party lines expressed the following concern: "The Pepsi-McDonald's invasion . . . comes close on the heels of unchecked entry by the multinational corporations into the banking system, stock markets, telecom services, the markets for alcoholic and nonalcoholic beverages, snack foods etc. . . . The entry of U.S.-style junk foods into India is particularly reprehensible for its deleterious likely fallout on ecology, public health and economy."[34]

Leading the attack against the "Pepsi-McDonald's invasion" from the radical right wing of the political spectrum was the Swadeshi Jagran Manch (SJM, or Self-Reliance Awakening Forum), a wing of Rashtriya Swayamsewak Sangh (RSS, or National Volunteer Corps). Traditionally, the RSS has been the most vocal advocate of Hindutva (Hindu essence) and is seen by many as the ideological spirit behind the powerful Hindu nationalist BJP. The RSS and the BJP are adamantly opposed to the Congress Party, which, they argue, has played a dangerous political game, for decades, of appeasing religious minorities, particularly Muslims, to gain their votes en masse. The Hindu nationalists blame the Congress Party and, more particularly, Mahatma Gandhi for agreeing to the partition of colonial India and the creation of the Islamic state of Pakistan as a necessary prelude to Indian independence from British rule in 1947.

The Congress Party, on its part, blames the Hindu nationalists for inflaming religious passions to gain votes of the majority Hindu community as a bloc and for creating dangerous rifts among India's diverse religious communities. The Congress Party does not forgive the BJP for continuing to ally with the RSS, some of whose members were implicated in the assassination of Mahatma Gandhi in 1948. However, in the overturned equation of economic liberalization in the 1990s, as the Congress Party embraced a transnational vision of consumption, it became increasingly distanced from its early vision of Gandhian nationalism. At the same time, Hindu nationalists strategically appropriated the Gandhian slogans of *swaraj* and *swadeshi* and recast them as rallying cries for the BJP's critique of the Congress Party policies of economic liberalization that have enabled transnational corporations to set shop in India.

With a statue of Mahatma Gandhi in the background, Hindu nationalists, led by the SJM, launched a campaign against Pepsi in the nation's capital, Delhi, demanding that the transnational corporation "Quit India." Also in the SJM campaign was George Fernandes, whose disenchantment with socialist-style critiques of the "Pepsi-McDonald's invasion" induced him to ally his left-wing Samata Party with the radical right-wing politics of Hindu nationalism. About 100 protesters who had gathered for the campaign smashed Pepsi bottles to the ground and shouted slogans against the economic liberalization policies of the Congress Party. "Foreigners go home," chanted the activists, as the campaign organizers passed out pamphlets entitled "Declaration of War." Protesters emptied the contents of bottles of Pepsi and burned a poster showing a Pepsi bottle topped by a hat bearing a U.S. flag. "We are targeting Pepsi first," the SJM activist Anook Agarwal declared. "But other products will follow like Coke, Colgate, Palmolive, Procter & Gamble."[35]

"Today is symbolic," Murli Manohar Joshi of the BJP proclaimed. "It is symbolic of the dangers to India and the need to protect India's political and economic independence."[36] The symbolic significance of SJM's anti-Pepsi campaign that Joshi refers to is that the demonstration coincided, to the day, with the fifty-third anniversary of the Quit India Movement that was led by Gandhi to overthrow British rule. The Gandhi memorial, where the SJM launched its protest against Pepsi, depicts Mahatma Gandhi leading protesters on the famous Salt March in 1930. As is well known, Gandhi's Salt March of 1930 was launched as a symbolic protest against British tax on native production of salt. As a mass movement, it captured the people's imagination and transformed the Indian nationalist movement into one of the most successful and nonviolent struggles against British imperialism.

"Gandhi Meet Pepsi." The irony of the equation becomes evident. As ideological compulsions quickly overwrite collective imaginations of nationalism and transnationalism, historical subtleties of political contestation in postcolonial India are reified into a symbolic competition between deified figureheads such as "Gandhi" and "Pepsi." There has always been a rich and diverse tradition of political critique and activism in India that seeks to hold political parties, the government, and transnational corporations accountable. Some of these activist groups and individuals have focused on lobbying against specific project proposals by transnational corporations like Cargill, DuPont, and Enron.

Other activist campaigns, like the well-known cases against Union Carbide and Nestle in the 1980s, have aimed to transform overall legislative and policy issues. In addition to lobbying and seeking to overhaul the legal and policy processes, many activist organizations and individuals in India target the misdemeanors of transnational corporations, national, state, and local governments, and political parties through staged rallies, demonstrations, sit-ins, blockades, public meetings, and street theater. Among the many groups and organizations are Jan Natya Manch and Nishant Natya Manch in Delhi, which regularly hold street plays. Other groups, like Azadi Bachao Andolan in Indore, print posters, pamphlets, leaflets, newsletters, and books in Hindi to disseminate information about transnational corporations that the national English-language media often do not provide. Research organizations, like the Public Interest Research Group (PIRG) in Delhi, provide the English-language press with exhaustive information about transnational corporations, detailed case studies, and in-depth critiques.[37]

In a telling study that was briefly reported in some sections of the national English-language press, the Public Interest Research Group reveals that contrary to governmental projections, Pepsi has created only 963 direct jobs in India. The study argues that Pepsi claimed the benefits of export quotas granted by the government by selling overseas products made by other companies.[38] However, Pepsi's version of the story is very different. Explaining that Pepsi created 30,000 new jobs and set an export target of 200 crore rupees (about $64 million) in 1995, Pepsi India spokesperson Deepak Jolly said:

> This does not count the 2,000 people directly hired in Pepsi's Indian plants. . . . Pepsi counts the employees of its 16 bottlers, 11 franchisees and distributing agents as its own. . . . Pepsi International uses India as a base to supply glass bottles to ventures in Vietnam and West Asia and soft drink concentrate to Russia. Pepsi India exports tomato paste, basmati rice and has ties ups with local exporters to sell their produce overseas.[39]

In this confusing game of economic manipulations and political maneuvers, as critics of the government's economic liberalization policies—including academics, media persons, special interest groups, and social activists—voice concern over the looming vision of transnational consumption, they find themselves on the same side of the political spectrum as leftist groups like the Samata Party, who in turn find themselves on the same side as the right-wing groups like SJM, which brandishes slogans like "Computer chips not potato chips."

I. K. Gujral (who became the nation's prime minister in 1996) argues: "A new challenge is on us. I am not sure of those who feel that shutting out the world is the way. There is no going back."[40] Even the anticapitalists among Marxists are now forced to concede that there is no going back from the government's policies of economic liberalization. "The economic reforms must be more sophisticated," says Prakash Karat of the Communist Party of India-Marxist (CPI-M). "Potato chips are just the most visible manifestation of the flaws in the policy." Some writers in mainstream newspapers seem to advocate a more fatalistic approach to the seemingly unstoppable march of potato chips and cola drinks in the Indian markets. "Is our karma, cola?" wonders Jaideep Lahiri of the *Economic Times,* and he poses the following three rhetorical questions as "food for thought" to his readers: "One, why should politicians tell middle class consumers what they should eat and what they shouldn't? Two, if Pepsi is prevented from making potato chips, will people stop eating potato chips? More important, will those who can't afford potato chips get any richer as a result?"[41]

As the many flaws in the policy of economic liberalization were being played out in various media debates and political campaigns in 1996, even the most visible of its manifestations provided reason enough for the Indian electorate to reject the Congress Party, which suffered its worst defeat in history. However, in the historic general elections of 1996, no other single party could gain absolute majority in the Lok Sabha of the Indian Parliament either. Many saw this fragmentation of votes as the Indian electorate's will to refuse the contradictory ideologies of all the major political parties. The Congress Party suffered its worst electoral defeat ever, winning only 140 seats, while the BJP emerged as the single largest party, with 161 seats, and the remaining seats were captured by socialists, communists, and several regional parties.

A minority government appeared inevitable, and the BJP staked its claim as the largest party in Parliament, with the hope that, once in power, it could muster support from other smaller parties. In an attempt to convince skeptics of the Hindu nationalists' ability to respect the religious diversity of the nation, the BJP selected Atal Behari Vajpayee, the most moderate of its national leaders, as its candidate for prime minister. The BJP also sent a clear message around the world to transnational corporations about its intent to continue the economic policies of the previous government by giving the key Finance Ministry to Jaswant Singh, the most articulate exponent of economic liberalism in the Hindu nationalist front. However, the BJP government lasted only twelve days, as a new coalition of regional parties, socialists, and communists hurriedly joined forces under the name of a "United Front" to stake their claim to power.

As the United Front, led by the socialist Janata Dal, did not have the required majority in Parliament, the Congress Party was constrained to support a coalition of its staunch critics, with whom it had no electoral alliance. The irony of the situation was all too obvious, for these parties fought each other bitterly before, during, and even after the general elections. The only thing that obliterated their political differences was their unified opposition to the Hindutva brand of nationalism preached and practiced by the BJP and its allies. After hectic parleys, the United Front came up with a "Common Minimum Program" of electoral promises as a noncontroversial compromise that was acceptable to all the constituents of the coalition. The Common Minimum Program also appeased the Congress Party, which had the power to pull the rug out from under the government anytime it pleased, by withdrawing its support for the United Front in Parliament.

The "Common Minimum Program" was presented to the nation by the United Front leadership as a manifesto of the coalition's commitment to socialist principles of poverty alleviation and equitable national development. Yet the United Front government gave the key portfolio of the Finance Ministry to P. Chidambaram, who had earlier resigned from the Congress Party over differences about, among other things, the slowing pace of economic liberalization, and launched a regional party called Tamil Manila Congress (TMC) in his home state of Tamil Nadu. By appointing Chidambaram to the key portfolio of the Finance Ministry, the United Front government sent a clear signal to the Congress Party and to transnational corporations alike that the economic liberalization of the Indian markets would continue unabated.

However, even before the United Front government could settle in, the problematic of power that had dogged the preceding governments began to rear its uncompromising head. To govern, the United Front had to convince the Congress Party that the latter's policies of economic liberalization would not be derailed. By allying with the Congress Party's liberalization policies, the left-wing constituents of the United Front—consisting of socialists, communists, Marxists, and some regional parties—feared that in the absence of a strong leftist agenda, a large segment of the electorate that was disillusioned with the Congress Party would be lured away by the BJP and its radical fringe, which preaches xenophobic nationalism.

To counter both the radical rhetoric of the right wing and the centrist ideology of the Congress Party, the United Front had to proclaim itself as a unique leftist/socialist formation, embracing the policies of economic liberalization as a means to attain the coalition's stated goals of poverty alleviation and equitable development in the national community. Yet this created the danger of losing the confidence of powerful segments of the middle class—the business and industry leaders who support greater liberalization of the Indian markets.

Mired in the matrix of power, the United Front, led by the socialist Janata Dal, found itself run over by a juggernaut of its own making, as it attempted to bring together the policies of economic liberalization and socialist development, while trying also to sustain a political ideology that would make it a unique formation distinct from both the Congress Party and the BJP. Irony quickly turns into parody as those at the helm of governance are forced to overwrite their political ideologies to death in order to maintain their hegemony in the transformations of nationalism and electronic capitalism in postcolonial India.

Protests and Politics in Electronic Capitalism

The peculiar problems and potential of articulating the political power of nationalism to the emerging order of electronic capitalism became particularly clear in November 1996, when the Miss World competition was held for the first time in India, in the picturesque "Garden City" of Bangalore. As the politics of nationalism, transnationalism, and translocalism literally collided in the streets of Bangalore, it became clear how many of the cultural tensions between the old and the new, tradition and modernity, left-wing and right-wing, patriarchal ideologies and feminist critiques were turned topsy-turvy by the rapid spread of electronic capitalism in postcolonial India. Promoted as a "tribute to Indian culture," the Miss World pageant was set against the backdrop of reconstructed ruins of a fourteenth-century Hindu temple inside a refurbished cricket stadium in Bangalore.

Eighty-eight women from around the globe competing for the coveted Miss World title wore long, transparent skirts around their swimsuits in what the organizers described as a mark of their deference to Indian cultural traditions. Hosted by Amitabh Bachchan, the Indian film industry's biggest box-office attraction for over three decades, the "tribute" was an elaborate three-hour extravaganza with hundreds of performers, dozens of elephants, and pervasive appropriations of traditional Indian music, apparel, and cultural iconography.

With an estimated 20,000 people in attendance, the gala event furthermore proved its appeal among Asia's elite. Filling the front rows of the venue were Bangalore's rich and famous, as well as movie stars, celebrities, politicians, and international personalities like the sultan of Brunei, who reportedly bought 200 tickets for his entourage at $695 each. Less fortunate spectators had to be content with watching the event live on national television, along with an estimated transnational audience of two billion viewers.[42]

Outside the stadium in Bangalore, thousands of protesters had assembled, constituting more than "a dozen Indian groups, including feminists, communists and Hindu politicians," who, according to CNN's correspondent

Anita Pratap, had "threatened to storm the venue and disrupt the show."[43] She also reported that a feminist activist and fourteen other women had threatened to sneak into the pageant and burn themselves in protest. Noting that "there were bomb threats as well," Pratap pointed out that "these groups opposed the beauty pageant alleging it demeans women and corrupts Indian culture."[44] Beginning with only a few demonstrators before noon, the crowds swelled continuously throughout the afternoon, and at 4:00 P.M. around 1,000 protestors were arrested as they tried to march on the stadium. By the evening, Pratap reported that "the situation in Bangalore was quite tense" and the city had become a "theater of street demonstrations, protest marches, mock pageants, litigations and effigy burnings as protestors tried to block the show."[45]

Security at the venue was very tight, but just before the show began, violence erupted as a group of Hindu activists attempted to cross a road blockade. When they began throwing rocks, police used batons and eventually fired tear gas and rubber bullets to quell the protest. Authorities reportedly arrested 1,300 protesters, and another fifty sought medical treatment for injuries sustained during the demonstrations. Although protests against similar beauty contests are common in many parts of the world, few have resulted in such heated confrontations or have attracted the exceptional amount of media attention that the Miss World pageant in Bangalore garnered.

Like many other postcolonial nations, India's fascination with beauty contests seems to be connected to transformations wrought by the emerging culture of a transnational consumption. The expansion of satellite television services and the concerted government policy of economic liberalization since 1991 are often cited as the forces fueling widespread changes in the national community. However, even in past decades, a large number of beauty competitions were staged with considerable pomp and pageantry at exclusive clubs, women's colleges, and even high schools throughout the country.

At the national level, *Femina,* India's leading women's magazine, organizes an annual Miss India contest whose winners go on to participate in international competitions like Miss World and Miss Universe. These events have drawn sporadic and localized protests from political groups and women's activists, who find beauty contests "unhealthy and sexist" and "derogatory to women in general."[46] At the same time, beauty contests on the global stage have inspired significant outpourings of national pride, as when Miss India, Sushmita Sen, won the Miss Universe title in Manila in 1994. National enthusiasm grew even more intense when, later that year, India's Aishwarya Rai was crowned Miss World in Sun City, South Africa. In both cases, numerous local protests were drowned out by a chorus of patriotic celebration that hailed this peculiar play of fate in the lives of two young women as an omen of India's rise to global prominence.

The *Hindustan Times,* a widely circulated English daily, captured this sentiment when it announced Rai's stunning victory on its front page with a bold headline playing on the popular Onida TV commercial slogan "World's envy, India's pride."[47] After Rai's victory as Miss World, an overjoyed Sen exclaimed, "We have conquered the world," an assessment that seemed to encapsulate this outpouring of nationalist fervor.[48] The terms of this conquest were later explained by Sathya Saran, the editor of *Femina;* India, she said, "is now more receptive to and more aware of the international look. We have adapted ourselves over the years and are now in tune with international standards."[49]

Vimala Patil, a former editor of *Femina,* claimed that India's stunning feat on the global beauty stage was possible not only because "Indian girls ... are better prepared but because India has been in the eyes of the world thanks to its economic reforms."[50] This explicit link between beauty and business is often made when analyzing India's recent rise to prominence on the global stage. It is frequently pointed out that transnational sponsors are now flocking to beauty contests on the Asian subcontinent in hopes of tapping into reputedly vast and growing Indian consumer markets.[51] Dispensing beauty titles is a relatively inexpensive way to build brand recognition among the country's consumer population, which is estimated to be somewhere between 150 and 485 million people, depending on the type of product being marketed.[52]

Calling India the "world's largest emerging market," Noel V. Lateef, president of the Foreign Policy Association in New York, writes: "The pace at which India has adjusted to the transformative world market has surprised even the most cynical observers. Having made the hard decisions to reform its economy, India is easily the most significant test case in the world for whether democracy and capitalism can triumph over mass poverty."[53] Consequently, many have commented that the attractions of India's emerging consumer markets go hand in hand with the success of Indian women on the global beauty stage. Yet the connections between these trajectories is obviously more subtle and complex in the transnational order of electronic capitalism.

The beauty contests not only *represent* the power politics of electronic capitalism in India but also are *marketable* events for the nation's new image in the transnational order. Over the years, many participants in beauty pageants, like Zeenat Aman in the 1970s and Juhi Chawla in the 1980s, have leveraged their exposure at these events into highly successful careers in the Indian film industry. More recently, when Sen and Rai won the Miss Universe and Miss World crowns, respectively, in 1994, they were flooded with offers from celebrated directors and producers, not to mention other lucrative opportunities like modeling and product endorsements.

Not surprisingly, then, "the beauty business," as Jain puts it, "is an all-pervasive phenomenon" in India that "starts off with a Miss Beautiful contest in High school, goes on to chick-charts for the 10 most beautiful women in college and ends up at the Miss India extravaganzas which bring in fame, money and glamour."[54] Thus, Jain finds that "now every girl worth her Barbie doll has extended her horizons to the Miss Universe and Miss World pageants."[55] Without falling prey to exaggeration, one must recognize that the success stories of contestants at the Miss India and more recently the Miss World and Miss Universe pageants have been few and far between.

For every woman who triumphs at beauty pageants and rises to stardom in Indian films, there are millions of women whose dream is restricted to vicarious experience via tabloids and television. Consequently, the pervasive impact of this beauty economy is most crucially attached to the media imagery it produces. Yet this same media imagery, which is so significant in fueling the consuming desires of aspiring young women, is also passionately disturbing to other elements within Indian society who seek to resist the growing influence of the beauty business.

In this context, one can find many reasons for the extensive media attention to and the intense protest against the Miss World pageant in India. Certainly one can ascribe the media attention and the protests to the contentious debates over the government's policies of economic liberalization, accompanied by the explosive growth of satellite television, which promises to deliver the latest fads and fashions of transnational consumer culture into homes across the nation. One can also explain the passionate conflict in terms of the acute ambivalence toward the visible excesses of transnational consumption that electronic capitalism has produced in a traditionally austere national community. Indeed, many women's groups were concerned not simply about the pageant itself but about the general relationship between media imagery and the overall status of women in India. One of the leading protest groups had, for example, been trying to pressure the Indian film industry to temper the often-lurid representations of women in commercial cinema.[56]

On the face of it, many of the protests by women's groups are about the exposure and exploitation of female bodies for a masculine gaze. Bollywood filmmakers had for years built their success on narratives structured around climactic dance numbers featuring a beautiful young woman performing for the pleasure of male audiences. Moreover, since economic liberalization and the introduction of satellite television in 1991, feminist groups have confronted an avalanche of increasingly lurid images of women performing for male spectators in television and advertising as well.

Protesters characterized the Miss World contest as yet one more example of the subordination of women to the status of sex objects in the beauty business and on the television screen. On one level, these sentiments were tied to the simple fact that economic liberalization is not delivering the widespread prosperity promised by government leaders. At another level, the widespread discontent was being manipulated and harnessed by various political parties in anticipation of a midterm poll, if and when the shaky coalition government led by Prime Minister Deve Gowda collapsed.

As an editorial in *Asia Week* points out, several political parties were blatantly opportunistic in exploiting the Miss World protests to gain leverage in India's always contentious electoral politics. Among the political formations involved in such a maneuver were the left-wing communist parties in India. Although they were partners in the coalition government led by Gowda, left-wing parties like the Communist Party of India (CPI) and the Communist Party of India-Marxist (CPI-M) were vociferous in their criticisms, despite the obvious embarrassment it caused for the prime minister, in whose home state the pageant was being held.

Mr. M. D. Nanjundaswamy, a veteran socialist and the leader of a farmers' group opposed to the presence of transnational corporations in his home state of Karnataka, described the Miss World show as "just another instance of how India is falling for the worldwide trend toward commoditization and internationalization of everything."[57] Yet the ironic reality of the situation was that the leftist and socialist parties, as members of the coalition government, were party to the decision to allow the Miss World pageant to be held in India in the first place. Nevertheless, their critiques of Miss World seemed to strike a resonant chord among the large blocs of voters who tend to treat the power of transnational corporations like Pepsi and KFC with hostility and to view the excesses of consumer culture with great suspicion.

An extensive national opinion poll launched some months after the controversial pageant revealed that only 29 percent of the nation's population felt that it had benefited from the economic reforms since 1991. For most of the electorate, the rhetoric of globalization was not resonating with their personal lives, and many leftist critiques of the Miss World contest sought to build on this strain of discontent.[58] As the scale of the preparations in

Bangalore grew ever more bloated, critics furthermore emphasized the glaring contrast between the elitist excesses of the media extravaganza and the stark reality of a country where 200–300 million people still live in abject poverty.

At the time of the Miss World contest, the Hindutva movement led by the BJP was maneuvering to win the confidence of the majority Hindu community by drawing attention to the growing discontent with the government's professed policies of economic liberalization. In the short term, the BJP sought to fuel the cultural anxieties of nationalism by proclaiming its opposition to the Miss World contest, but in the long run it was also building upon the nativist claims of Hindutva that lie at the core of its political formation. In doing so, the BJP and its Hindutva allies generated a powerful movement against the Miss World pageant that had the potential to appropriate all other forms of cultural criticism—based on the politics of gender or locality—in the national community.

For instance, among the most contentious issues in the Miss World pageant and one that allowed observers to conflate the concerns of the feminist movement with the protests of the Hindu nationalists was the swimsuit competition. Exposure of women's thighs and ankles in the swimsuit competition was seen by many in the feminist movement as yet another example of the objectification of the female body for the male gaze. To the advocates of Hindutva, the swimsuit competition was an assault on traditional mores that was considered all the more atrocious because of the fact that it was orchestrated by foreigners, led by the London-based organizers of the Miss World contest. The event was characterized as a violation of Indian tradition, which the BJP and its ideological allies have been keen to portray as distinctly Hindu in character.

Protest groups were successful at pressuring the Miss World organizers to relocate part of the contest to the Seychelles, where contestants were jetted in to disrobe for the required swimsuit competition. Political criticism of the swimsuit competition also was articulated by the Hindu nationalists in relation to a variety of concerns about the policies of economic liberalization that fueled anxieties about appropriate standards of female sexuality in India. From the Hindu nationalists' perspective, the beauty pageant's apparent assault on traditional Hindu/Indian norms not only gave legitimacy to the passionate mass response but justified the participation and even leadership of women in the protest movement.

Moreover, the publicly expressed indignation of women's groups only helped to *legitimize* the protest movement, although that indignation, as I will explain, had to be channeled and contained within the ideological confines of the Hindu nationalist movement. The feminist groups could "name" the offense to the dignity of Indian women but ultimately could not claim the struggle over its meaning as their own. Commenting on this ideological tension between the politics of gender and nationalism in colonial and postcolonial India, Kum Kum Sangari writes: "women must never name the social relation they are trying to preserve [and they must not] present it as a personal or material interest; they can only name the abstraction—family, honor, religion, nation—to which the social relation is either directly attached or which mediates it. . . . In the naming and the not naming resides the distinction between villainous and heroic inciting women."[59]

In their campaign against the Miss World pageant, Hindu nationalists invoked abstract notions of nation, family, and honor to create a tactical alliance with the leaders of the feminist protest movement in the streets of Bangalore. Many of the women's groups protesting the Miss World pageant were not unaware of the tradeoffs involved in the strategic alliance with the politics of Hindu nationalism advocated by the BJP and its supporters. As one of the feminist leaders, Brinda Karat, put it: "We consider this slogan of Indian culture a euphemism for reinforcing the fundamentalist viewpoint of woman as subordinate. In the name of Indian culture, what they really want to project is a stereotyped image of the meek and submissive Indian woman."[60]

Despite such cultural contradictions, leaders of women's groups—rather than any of the political parties—took the lead in challenging the Miss World pageant. For instance, the Mahila Jagran Samiti (Forum for Awakening Women), petitioned the Karnataka State High Court to prevent the staging of the Miss World pageant on grounds ranging from cultural sovereignty to national security to public health. The lower court judge who initially heard the case dismissed the petition, but the High Court sustained some of the objections during an appeal and put a number of restrictions on the staging of the event. One of the restrictions was that the Indian organizer of the pageant, Amitabh Bachchan Corporation Limited (ABCL), was prohibited from selling alcohol during the event and from holding the controversial "bikini show." It also had to submit to court oversight its security arrangements for the event. Despite all these concessions, the economic stakes for ABCL were high enough for the company to comply with the High Court's ruling. With that, all the legal hurdles were cleared, and the High Court overturned the lower court's decision, thus giving the organizers a green light for staging the contest in Bangalore, with restrictions.[61]

The High Court's decision was a major blow for the women's groups who had filed the case to protest the Miss World pageant. When the recourse to legal action provided only limited success, a group of women's activists, led by Kina Narayana Sashikala at Mahila Jagaran Samiti, announced that opponents would pursue alternative tactics. Among them, Sashikala was quoted as saying, would be a dramatic action: "We will sneak into the stadium and burn ourselves. We already have the tickets."[62] This widely reported plan alarmed pageant organizers, since only a week earlier a twenty-four-year-old tailor in the south Indian city of Madurai had doused himself with petrol and set himself alight, apparently in protest against the Miss World contest.

Expressing concern at the threats of mass suicide, Julia Morley, the chairwoman of Miss World Ltd., said: "I hope [the protesters] will act as women and come and talk to us instead of acting as rebels." Shifting attention to the contestants, Morley told journalists: "Let us have respect for them. There are 89 of them and they are young and beautiful. Let us take care of them."[63] In four brief sentences, Morley sought to universalize the contests' standards of femininity by suggesting that women are nurturing, rational, communicative, and of course young and beautiful. But it is the power relation implied by the invitation Morley extended from behind the battlements of Bangalore—let *them* come to us—that highlights the very frustrations confronted by the opponents of Miss World.

Allied with regressive political elements and struggling for a cosmopolitan standard of women's rights, some women's groups found themselves positioned as marginal fanatics challenging a rule-governed competition legitimized by a supposed global audience of two billion. In the eyes of many who were following the events in Bangalore, women's groups had, in desperation, abandoned the high ground. Responding to the expressed concerns of the organizers, the state police quickly ordered Sashikala's arrest. Under Indian law, the police can arrest a person if there are reasonable grounds to suspect the intent to commit suicide. Faced with the possibility of imprisonment, Sashikala went into hiding, which severely curtailed her protest activities.

In order to remain involved in the movement, Sashikala sought the help of BJP activists, who claimed that they were constantly in touch with her during the finale of the Miss World pageant.[64] Physically invisible and now communicating in public through the channels of BJP activists, Sashikala's predicament symbolized the Hindutva movement's political and discursive absorption of the feminist challenge to the Miss World contest.

The predicament faced by Sashikala in the Miss World controversy brings to attention some of the many problems faced by women's groups when they strategically align with Hindu nationalist parties like the BJP to ensure the momentum of their protest movement against the growing hegemony of electronic capitalism in India. In this strategic alliance, what were initially cast as women's issues became increasingly associated with popular resentments against economic liberalization and transnational corporate influences. Instead of giving voice to the feminist critiques of the Miss World contest, public deliberation and media coverage increasingly focused on BJP activists' characterizations of the struggle as an attempt to defend a distinctively *Hindu/Indian* concept of femininity against the profane values of invading of "foreigners."

The BJP activists' Hindutva critique of Western notions of femininity thus emerged as the most prominent rationale for opposing the Miss World pageant. Ironically, the women's groups that had been at the front lines of the protest movement in the streets of Bangalore were quickly marginalized as minor players in the debate, and the Hindu nationalist critiques of female sexuality in the Miss World contest became the symbolic condensation of what India had to fear most from the government's policies of economic liberalization and the growing power of transnational corporations.

A few months after the Miss World pageant concluded in his home state of Karnataka, Prime Minister Gowda was forced to resign from office when the Congress Party strategically withdrew its support, with an eye toward the midterm elections, when it hoped to benefit from the electorate's disenchantment with the United Front government. Although the Miss World controversy had little to do directly with causing the fall of the coalition government, it certainly did bring into relief the ideological quandaries of the various political parties as they struggled to perform a delicate tightrope walk between the contending pulls of nationalism and transnationalism in postcolonial India. The power politics between the various ideological formations were once again prominently displayed on national, transnational, and translocal networks during the general elections of 1998—considered by many to be India's first television elections.

Overwriting the Nation on Television

The elections of 1998 were a very different phenomenon from general elections in the past where Doordarshan had been the only network providing extensive coverage of the battle for the ballot. In the general elections of 1996, many of the foreign and domestic satellite networks were unable to provide live coverage because of the restrictions the government put in place on access to uplinking facilities from within the country. However, during the general elections of 1998, the government relaxed its rules on satellite uplinks for eight weeks, enabling private media networks, like the BBC, CNN, Star TV, and Zee TV, to provide round-the-clock news coverage for the first time in the history of Indian television.

In south India, regional-language networks, like Sun TV, Vijaya TV, ETV, and Asianet, also focused more attention on the general elections and increased the number of news bulletins in their daily schedules. While the state-sponsored network, Doordarshan, has been known to tilt its news coverage in favor of the party holding the reins of power in the central government in Delhi, the regional-language channels Sun TV, Vijaya TV, ETV, and Asianet started a new trend of picking favorites on the basis of their relationships with the political parties at the helm of affairs in the state governments, particularly in South India, where the regional-language channels attract large audiences.

In a country where the literacy rate is no more than 52 percent, the power of print media to influence the political agenda has historically been quite limited. Citing a study conducted by the Center for Media Studies, Mannika Chopra argues that "while 32 percent of the people in a rural pocket of Andhra Pradesh, a state in south India, had heard of Sonia Gandhi through the medium of newspapers, as many as 49 percent had heard of her through television."[65] Surveying expert comments about the potential impact of television on grassroots politics in the general elections of 1998, Chopra writes: "The pundits spoke of the beginning of an era of tele-democracy—a time when the small screen would substitute completely for a candidate's guts-and-gore speech. They pointed to how villagers near the small town of Betul were suddenly speaking of 'vote swings' and 'incumbency factors' using vintage TV pollster jargon."[66]

Despite the numbers of foreign and domestic networks providing extensive news of the elections, television coverage across the channels emphasized sound bytes over substantive discussions and critical analyses. Prannoy Roy, India's best known television pollster, conceded that politics had become more like sports. "It's all about who wins, who loses, not about issues," he declared.[67] As election results trickled into newsrooms, television anchors and election analysts provided live commentary and displayed detailed scorecards, using a wide variety of graphics and special effects that are characteristic of major sporting events like cricket in India.

In a media campaign that resembled a quiz show or a game show on television rather than a political debate on electoral strategies, the Congress Party released a list of five questions to news reporters that, the party spokesperson V. N. Gadgil claimed, were directed at Atal Behari Vajpayee, who was the BJP's declared candidate for the prime minister's post. The five questions were as follows.

1. Article 370 was introduced in the Constitution of India when your leader, Shyam Prasad Mukherjee[,] was a member of the Nehru cabinet and he agreed to the introduction of Article 370. Do you disown your leader Shyam Prakash Mukherjee?
2. You have an alliance with Akali Dal. Do you support the Anandpur Saheb Resolution? "
3. Your spiritual Guru Golvalkar had in a famous interview to a Marathi Daily *Navakal* defended and justified *Chaturvarna,* the caste system, which he said is essential for the survival of Hinduism. Do you agree?
4. Do you deny that in 1942 you apologised before a magistrate for your participation in the 1942 Quit India Movement?
5. Do you accept Mahatma Gandhi as the Father of the Nation?[68]

In releasing these questions in the form of a pop quiz to the media, the leaders of the Congress Party hoped to raise questions not only about the personal integrity of the BJP's prime ministerial candidate but also about the Hindu nationalists' ability to articulate a coherent political ideology of secular nationalism. Speaking on a television show, Gulam Nabi Azad of the Congress Party made the following observation, perhaps partly in ridicule and partly in frustration: "The BJP has claimed our slogans, manifesto, leaders and even our men. Issues such as secularism, stability, *swadeshi,* leaders like Subash Chandra Bose, Mahatma Gandhi . . . the BJP has nothing of its own."[69] Margaret Alva, a spokesperson for the Congress Party, argued that the BJP was entering into "alliances of convenience" with parties "of different hues" that did not share its political ideology in an obvious attempt to grab power at any cost.[70]

When the votes were finally tallied, the result was what many pollsters had predicted and all the political parties had feared before the elections: a hung Parliament, with no single party gaining majority in the Lok Sabha to form a stable government. For those who had tuned in to the live coverage of the elections on television, the game had ended in a very familiar anticlimax—a draw. The BJP once again emerged as the single largest party, with 179 seats, and the Congress Party remained in a distant second place, with 145 seats.

Although the BJP and its electoral allies managed to garner 250 seats, they were still woefully short of the simple majority required to gain control in the 543–seat Lok Sabha.

In a nationally televised vote-of-confidence debate, the new government, led by the BJP, attempted to articulate a stable political identity for its coalition that would win over not only of those in Parliament but also viewers across the nation who were watching the event live on television. However, the dilemmas of power-sharing that had dragged both the Congress Party and the United Front to great falls in past years soon began to dog the Hindu nationalists. To win the confidence of the nation about its ability to rule reasonably, the BJP was forced to project a moderate identity for its coalition government.

However, this meant running the political risk of alienating the party faithful, who are drawn to the radical fringes of the Hindu nationalist movement. To retain the confidence of the radicals, the BJP had to openly embrace the xenophobic elements of the Hindutva ideology that the party claims as its founding principles. Yet this created the danger of losing the confidence of powerful segments of the urban middle class, the business and industry leaders who support greater economic liberalization in the national community. Caught in the vortex of power, the BJP found itself run over by a juggernaut of its own making as it attempted to juggle its policies of economic liberalization between the contradictory pulls of nationalism and transnationalism, even as it tried to sustain the Hindutva ideology that had enabled its meteoric rise in the political arena.

As the nationally televised debate in Parliament revealed one political contradiction after another, opponents of the BJP-led coalition could barely conceal their glee at the quandary of the Hindu nationalists. All major political parties—centrist, leftist, regional—refused to align with the BJP until the party revealed its "true" identity. Faced with the unimaginable moment of truth in power, the BJP was both unable and unwilling to articulate a stable identity to the satisfaction of its political opponents inside and outside Parliament. Not having the required majority in Parliament, the BJP-led government was forced to resign, after thirteen months in power, when a key coalition partner, the AIADMK, withdrew its support.

Another round of midterm elections were held in September–October 1999, and the BJP once again emerged as the single largest party but without the required majority to form a government on its own. This time, however, the party managed to form a more stable coalition, with 298 seats in the Lok Sabha, while the Congress Party and its allies remained a distant second, with 135 seats. Other parties that had no political truck with either the BJP or the Congress Party garnered a respectable 104 seats. In the short span between 1996 and 1999, an unprecedented three general elections were held, in which no single political party was able to gain a simple majority in the Lok Sabha, and no single ideological formation was able to capture the collective imagination of the Indian electorate.

With even the most powerful in the political arena appearing powerless in the context of the rapid transformations of nationalism and transnationalism in postcolonial India, many Indians have taken recourse to writing about a sense of community in the past tense that, in contrast, appears unchanging. Suggesting that there is now a complete absence of leadership in the national community, the noted legal expert Nani Palkhivala opines: "If India has ever produced a leader, it was Mahatma Gandhi."[71] Others, like Butalia, who find that Gandhi's name is now being overwritten by transnational corporations like Pepsi, tell the readers of the *Independent* about the changes wrought by the Pepsi brand of consumer culture at the college where she teaches in Delhi. There, like many other colleges in cities across India, students come "in their air-conditioned cars, stereos blaring, jeans and Benetton T-shirts to the fore."[72] The cheapest of these cars, Butalia tells us, costs more than three years' salary for a university lecturer.

Contrast that with college life in the seventies, she recounts, when students like her rode on buses with an "all route-bus pass [costing] the equivalent of 40 p."[73] Butalia recalls the days, not very long ago, when most women students would wear the traditional *salwar kameez* to college. Today, she tells us, students come in miniskirts, off-the-shoulder T-shirts, and tight dresses "and show a lot of leg and thigh and even a little bosom." There is a whole generation that still fondly remembers a collective sense of community, as Butalia does, during the so-called good old days of Indian nationalism, when

> the *subziwallah* came to your door every morning with his cart offering fresh vegetables; the small grocery store down the road stocked Indian brand names; an imported refrigerator was unusual. But now the supermarket ("one-stop shopping—so you can shop like housewives the world over") is replacing the *subziwallah*. And you can buy refrigerators made by Sanyo, faxes by Canon and Minolta, stereos by Sony, computers by Apple and Olivetti. If you go abroad, no one needs ask you [any more] to bring back Levis or Wranglers. Even the roadsweeper has abandoned his *pyjama-kurta* for baggy jeans.[74]

Butalia turns to a philosophically inclined friend who reminds her that "[f]ood, women's clothes, and rituals relating to birth, marriage and death

... are always the last to change" in the vortex of cultural flux. But now, with everything changing, Butalia argues, "Death, perhaps, is the only constant."[75] There is a fair amount of philosophizing and romanticizing going on in India today. A lot of politicizing too. Once it is in the realm of politics, even the philosophical constant "death" no longer remains unchanging, assuming a symbolic life far more powerful than anything even the most powerful may have to offer. In this consuming domain of power, as proponents and opponents of competing visions of nationalism and transnationalism are forced to overwrite their own identities in India, they are, as Butalia puts it, "undermined either by their allies or by their own humbug."[76] The only way to elude the humbug, she suggests, is to refuse the consuming power of electronic capitalism that foreign and domestic networks bring into the national community of Indian television.

However, Butalia also recognizes that to refuse the desire for consumption is "to eschew television, cars, air-conditioners, computers," and to embrace a life of "renunciation of luxury and a living out of your beliefs."[77] She acknowledges the daunting task of living such a life in our times, and suggests that it would require all Indians to be like the Father of their Nation, Mahatma Gandhi, who symbolizes the life of renunciation, in contrast to the Pepsi brand of transnational consumption. Alas, Butalia concludes with apparent dismay, "there are very few Gandhians now."[78] A cry of despair: The Father of the Nation is dead! The nation is dead! Its author has disappeared! There is no sense of community anymore in India today!

Is There an Indian Community of Television?

"Gandhi Meet Pepsi." Now apparently silenced even in the minds of his most ardent admirers by the consuming power of electronic capitalism, the Father of the Nation, Mahatma Gandhi, is thus overwritten to death in a symbolic exchange with transnational corporations like Pepsi. In this process of overwriting, all that seems to remain of the national sense of community is the spectral figure of its author, now shorn of all its significance, save for its titular symbolism as the Father of the Nation. However, as the contentious discourse of economic liberalization in India reveals, even as a specter Gandhi continues to be a politically significant figure, whose name is constantly invoked as a grand author of the national community, as it negotiates the political vagaries of economic liberalization.

Obviously, then, it is not sufficient to merely create catchy headlines—such as "Gandhi Meet Pepsi"—that announce the death of the national community and trumpet the birth of a new transnational formation of electronic capitalism. In this context, it becomes just as insufficient to proclaim hollow slogans—telling Pepsi to "Quit India"—in an attempt to create a romanticized image of the national community by artificially resurrecting historical narratives of the great struggles of the Gandhian movement of the past. In raising the question "Is there an Indian *community* of television?" in the emerging order of electronic capitalism, I argue that it is futile to merely celebrate, denounce, or despair about the state of nationalism. Instead, I argue, we must be willing to address the political significance of the possibility of the very disappearance of the nation as an "imagined community."

By deconstructing the ideological constructions of nationalism and transnationalism in electronic capitalism, I have attempted to reveal how the Father of the Nation remains as a spectral figure who is symbolically overwritten in the empty spaces left in the wake of slogans proclaiming his death. The process of overwriting-to-death that I refer to is a paradoxical dual process of writing and erasure—as is encapsulated in the headline "Gandhi Meet Pepsi." In this creative headline, the invitation to India, as a national community, to embrace a new vision of electronic capitalism is written in the name of the Father of the Nation, who is symbolically overwritten by transnational corporations like Pepsi. However, the flip side of the story reveals that even as transnational corporations like Pepsi seek to expand the global reach of electronic capitalism (through television), national governments seek to exploit changes wrought by economic liberalization to sustain the legitimacy of their nationalist visions (through television).

When transnational and national (tele)visions collide and collude, they invariably make room for translocal (tele)visions of community that undermine both the transnationalist *and* the nationalist quest for hegemony. In this struggle for hegemony, what gets fleetingly revealed is the specter of a Gandhian vision—a spectral television if you will. The script for this silent television is written not by a struggling subaltern author who, as Gayatri Spivak points out, has been silenced. Rather, I argue that the script for this spectral (tele)vision is silently inscribed in the name of a grand author who is symbolically overwritten to death in the competing visions of nationalism and transnationalism in electronic capitalism.

To address the question "Is there an Indian *community* of television?" I turn to Michel Foucault's essay "What is an Author?" which provides us with an illuminating framework to discuss how the Father of the Nation remains as a spectral figure of an author underwriting the narrative transformations of nationalism and electronic capitalism.[79] Traditionally, in the context of a literary text and its narrative transformations in a community of readers, the status of an author, as an individual, raises several sociohistorical questions that, Foucault concedes, need to be acknowledged: How was the author individualized? What is the status of this author in relation to issues of authenticity and attribution? What are the systems of valorization in which this author was included? What was the moment when the stories of heroes gave way to the author's biography? What were the conditions that fostered the formulation of this fundamental category? These are sociohistorical questions that have been extensively studied in relation to the diffusion of what has come to be known as *auteur* theory in philosophy, literature, film, and, more recently, television studies.

Auteur theory gained considerable prominence in the 1950s and 1960s due largely to the works of a small group of French film critics and directors, among them most notably Francois Truffaut. Through a series of essays published in the influential journal *Cahiers du Cinema*, advocates of *auteur* theory developed a systematic framework to "find the artist in the art of the film." In this ambitious quest was embedded the desire to identify the creator of cultural productions by locating the author's unique imprint in individual films and as an entire collection of works. For proponents of art (generally noncommercial) film, like Andrew Sarris, the individual creator of cultural productions—the *auteur*—can be identified by locating the "inner meaning" and the "ultimate glory" that manifest themselves from film to film.[80]

For followers of commercial (usually Hollywood) film, like Peter Wollen, the *auteur's* identity in a film could be uncovered by looking for subtle elements of aesthetics and style that transcend the usually superficial treatment of fundamental and often irreconcilable motifs. The various attempts by critics and filmmakers to locate the identity of the *auteur* in "art" have been criticized for being too "formalist," "aesthetic," or "elitist." An additional criticism leveled against the proponents of *auteur* theory in commercial productions usually has to do with their seemingly uncritical acceptance of the "superficial," "coarse" "vulgarity" of the Hollywood culture. Many sociohistorical analyses of authorship in art and commercial films, as well as television shows, have revealed the remarkable cooperation between and cooptation of artistic and commercial considerations that is inherent in the production process of a filmic or televisual text.[81]

In this debate, Foucault's intervention is a remarkable attempt to depart from the conventional traditions of *auteur* theory in art and commercial media and focus instead on the sociohistorical transformations of authorship in a community. Foucault's writings suggest that it is now possible to address the rather complex questions of nationalist narratives in the transformations of electronic capitalism by restricting ourselves "to a singular relationship that holds between an author and a text, the manner in which a text apparently points to this figure who is outside and precedes it."[82]

To address the question of authorship in a Foucauldian framework, where the author appears as a figure who is both outside and before the text, the relationship between writing and death becomes a central concern. For writing in our time has inverted the age-old conception of authorship in folk narratives and religious epics, which were designed to guarantee the immortality of a hero. In the epic tradition, for instance, the hero could die an early death "because his life, consecrated and magnified by death, passed into immortality; and the narrative redeemed his acceptance of death."[83] In a different sense, this was the strategy for defeating death in the stories of *The Arabian Nights,* where storytellers "continued their narratives late into the night to forestall death and to delay the inevitable moment when everyone must fall silent."[84] However, Foucault finds that this strategy for using the spoken or the written narrative as a protection against death has been transformed in our times. Thus, writing is now

> linked to sacrifice and to the sacrifice of life itself; it is a voluntary obliteration of the self that does not require representation in books because it takes place in the everyday existence of the writer. Where a work had the duty of creating immortality it now attains the right to kill, to become the murder of its author. . . . In addition we find the link between writing and death manifested in the total effacement of the individual characteristics of the writer.[85]

Therefore, in order to comprehend the cultural significance of the simultaneous writing and erasure of the Father of the Nation as the author of nationalist narratives in electronic capitalism—as indicated by the headline "Gandhi Meet Pepsi"—we must "reexamine the empty space left behind by the author's disappearance" and pay close attention to the "new demarcations and the reapportionment" along the gaps and fissures of this void, even as we await "the fluid functions released by this disappearance."[86] In this fluid context of writing, an author may be designated by many names, may take many forms, may perform many functions. In postcolonial India, one such figure has been that of Mahatma Gandhi, who performs the designated authorial function of the Father of the Nation as he is overwritten to death by the competing visions of nationalism and electronic capitalism. After all, as Butalia seems to acknowledge in a disillusioned conclusion, the only good Indians are the few remaining Gandhians. Pepsi Meet Gandhi.

5 *Nikki Tonight,* Gandhi Today: Television, Glocalization, and National Identity

IN 1927, at the peak of India's freedom struggle against British colonialism, Catherine Mayo published her blatantly imperialist book *Mother India.*[1] Mayo's prejudiced view of Indian culture and traditions generated considerable controversy among the Indian literati, who called on the British government to ban the book. Mahatma Gandhi is said to have dismissed Mayo's book, describing it as "the report of a drain inspector sent out with the one purpose of opening and examining the drains of the country to be reported upon."[2] However, with characteristic irony Gandhi added that every Indian ought to read the book. There was something very—dare I say—postmodern about Mahatma Gandhi that defied rationalization and baffled his followers and critics alike. "My language is aphoristic, it lacks precision. It is, therefore, open to several interpretations," Gandhi once wrote to a confused follower. The reader will, therefore, I hope, understand my hesitation to explicate a rationale for Gandhi's paradoxical stance on Mayo's anti-India sermon, far less to elaborate a theory about his complex relationship with British colonialism and Western imperialism.

However, I would like to believe that in his paradoxical response to the Mayo episode, Gandhi was implying that one must be willing to use even the most prejudiced of criticism to interrogate one's own cultural straitjacket. As Ashis Nandy reminds us, Gandhi recognized that "the Mayos are transient phenomena; cultural renewal through internal criticism is a more serious, long-term affair."[3] It is with this Gandhian sense of cultural criticism that I seek to deconstruct a transient moment of unspeakable transgression of Mahatma Gandhi's name on the Star Plus Channel in India.

Star Plus Channel is part of Rupert Murdoch's STAR TV network, which beams programs into India and much of Asia from its base in Hong Kong. The moment of transgression on Star Plus, as I already mentioned, occurred on May 4, 1995, in an episode of the now-defunct talk show *Nikki Tonight*, when, chatting with the host, Nikki Bedi, a guest on the show, Ashok Row Kavi, a well-known gay rights activist and journalist in India, called Mahatma Gandhi "a bastard *bania.*" *Bania*, a Hindi word used to refer to a Hindu community of traders from the state of Gujarat in northwestern India from which Mahatma Gandhi came, is sometimes also used in a pejorative sense to call someone a miser.

The transgression was, quite literally, momentary. But soon it was engulfed in controversy, as the offended kin of Mahatma Gandhi filed a lawsuit in India, furious politicians, across party lines, demanded a ban on STAR TV, and angered audiences cried foul. The embarrassed mandarins of STAR TV, the producers of *Nikki Tonight*, Nikki Bedi, and Row Kavi all issued a public apology, and a few days later, the controversial talk show was quietly cancelled, "in deference and respect" toward India. In the final analysis, it becomes largely irrelevant whether Nikki Bedi and STAR TV failed to take Row Kavi's offensive remark seriously or whether they chose to ignore the remark to keep irreverent talk on the show flowing. For there were many in India who had taken notice of Row Kavi's effrontery on the talk show and were unwilling to ignore its transgressive significance, even if it was for a fleeting moment; or perhaps because it was for a fleeting moment.

In this chapter, I critically evaluate the controversy surrounding this fleeting moment of unimaginable transgression on *Nikki Tonight* by focusing on the crucial role that television plays in mediating collective imaginations of Indian nationalism by at once deifying and defiling the exalted status of nationalist icons like Mahatma Gandhi. By framing the debate over *Nikki Tonight* in terms of the question "Is there an Indian community of *television?*" I contend that the medium of television has emerged as the new battleground for competing visions of nationalism, transnationalism, and translocalism in postcolonial India. I conclude that *Nikki Tonight* is an exemplar text that reveals how collective imaginations of national identity and cultural differences have become ever more important, even as they are increasingly being blurred by the dynamic flows of electronic capitalism.

The Nation and Transgression

There is something very peculiar about momentary transgressions. A fleeting transgression, such as Row Kavi's insulting remark on *Nikki Tonight*, can be contained in what is apparently nothing of particular significance in the flow of conversation. And yet the moment of a fleeting transgression, such as the one on *Nikki Tonight,* can contain the potential for the most intense violence, rhetorical or otherwise. From Georges Bataille's extensive studies of transgressions we know that the violence of transgression is not organized violence as in war, nor is it animal violence: "It is violence still, used by a creature capable of reason (putting his knowledge to the service of violence for the time being)."[4] Although engendered by creatures of reason, there is nothing very reasonable or rational about transgressions.

For instance, it would be laughable to believe that calling someone a bastard would actually make that person's parentage dubious. Still, people continue to take offense when such insults are hurled at them or at their loved ones. Bataille hits it on the head when he says that taboos and transgressions have "a certain illogicality which makes it very difficult to discuss" their significance in logical terms.[5] Take, for instance, Row Kavi's abusive reference to Mahatma Gandhi on *Nikki Tonight.* On the face of it, there is very little of significance to discuss about this momentary transgression. As I have shown in the preceding chapters, Gandhi remains one of the most revered and abused figures in the history of Indian television, even more than half a century after his death.

So when Row Kavi called Gandhi a "bastard bania" on STAR TV, it wasn't the first time or the last time that someone abused the Father of the Nation in India. In fact, it wasn't even the first time that Row Kavi had called Gandhi names. On *Nikki Tonight,* when Nikki asked him about his career in journalism, Row Kavi recounted a fifteen-year-old incident: the English newspaper the *Times of India* had denied him a job because he had written a letter to the *Illustrated Weekly of India* in which he called Mahatma Gandhi a "bastard bania." When I met with Row Kavi at his apartment in the Santa Cruz area of Mumbai in August 1996, I asked him to briefly describe the two controversies involving him and the name of Mahatma Gandhi—the one on *Nikki Tonight* in May 1995, and the two-decades-old controversy with the *Weekly*—as he now remembered them.[6]

According to Row Kavi, sometime in 1971, when he was a trainee journalist with the *Times of India,* he wrote a letter to Kushwant Singh, the then editor of the *Illustrated Weekly of India.* In this letter to the editor, Row Kavi referred to Mahatma Gandhi as a "bastard *bania*," and it was later published in the *Weekly* with the words intact. After the publication of the letter, there was a furor, and many readers, politicians, and followers of Mahatma Gandhi criticized Row Kavi for his insulting remarks and the editor of the *Weekly,* Kushwant Singh, for publishing them. With tempers beginning to flare, both the editors of the *Weekly* and Row Kavi issued apologies, and the controversy died down as quickly as it had flared up. Shortly after that incident, over at the *Times of India,* where Row Kavi was a trainee journalist, it came time to consider him for a regular job. Row Kavi says that although he had passed at the top of his batch as a trainee, the then editors of the *Times* shot down his case because he had created a major controversy with his letter to the *Weekly.* On *Nikki Tonight,* when Nikki enquired about how he had started his career as a journalist with a controversy in the 1970s, Row Kavi recounted the fifteen-year-old story about his letter to the editor of the *Weekly.* He ended the story with the following comment: "You know instead of using those words, 'bastard bania,' if I had said that his mother was not Mrs. Gandhi when she gave birth to him, then I would have gotten away with it, and I would have got my job."[7]

Recalling the moment in later interviews with journalists, Row Kavi seemed to suggest that the political significance of his remarks did not seem to quite register on Nikki.[8] Other reports disseminating the news suggest that Nikki was taken aback momentarily, as she giggled uncomfortably for a few moments before continuing to chat away. Journalists covering the incident also reported that Nikki did not challenge the absurdity of Row Kavi's remarks or remind him that money did not mean much to the Mahatma. Instead, as Tim McGirk puts it, "the hostess sniggered like a schoolgirl on hearing a dirty joke."[9] As Nikki tried to laugh away the response, there was an uneasy moment of silence, but it was quickly overcome by a spate of other probing questions, to which Row Kavi gave other scandalous responses. For instance, among the other questions that elicited controversial responses from Row Kavi, there was one about who his favorite Hindi film actresses were. Row Kavi recalls the conversation thus:

> I said Madhubala and Rekha, out of whom, Madhubala, was straight. So then, immediately she [Nikki] turns to me and says, "does that mean Rekha is a lesbian?" Then I said, I didn't say it, you said it.... Then she asked me to name an actress I don't like, and I said Sharmila Tagore.... I did an imitation of Sharmila Tagore with her clipped convent [English] accent which represents a whole generation of falseness ... and the comment ended with a needless statement, but which is typical of the film industry, that most of these actresses' wigs are full of lice. You know this sort of lighthearted banter is going on. It was just a bitchy show. If you can't take it, then you better also sue every issue of *Star Dust* [the popular film gossip magazine in India].[10]

But apparently, not everyone was willing take his conversation with Nikki as "lighthearted banter" on "just a bitchy show," as Row Kavi puts it. In fact, there were some who took it very seriously and even decided to take matters into their own hands. For instance, Saïïf Ali Khan, the son of the actress Sharmila Tagore and the former Indian cricket star Mansur Ali Khan Pataudi, took offense at Row Kavi making fun of his mother on a television program watched by audiences in forty countries across Asia. According to news reports, Saïïf, who is a well-known Hindi film star, "forced his way into Row-Kavi's home and punched him several times.... Row-Kavi's mother, who attempted to shield her son from the angry actor, was also hurt."[11] A report in the *Times of India* quotes Row Kavi as saying "When my aged mother intervened, he hit her too."[12] According to the *Times* report, in the previous year, the Santa Cruz police station had registered a similar complaint against Saïïf and his wife Amrita Singh when they allegedly assaulted a female journalist.

Row Kavi, who lodged a police complaint against Saïïf for assault, maintains that the provocation for Saïïf's anger was something else, although he admits that his comments against Sharmila Tagore on *Nikki Tonight* could have been "the trigger." What motivated Saïïf to punch him, Row Kavi argues, was a review he had written for one of Saïïf's films, *Main Khiladi, Tu Anari* (I Am a Player, You Are Naive) in the April 1995 issue of *Bombay Dost,* an English-language magazine for Mumbai's gay and lesbian community in which Row Kavi's columns regularly appear.[13] One of the big hits of 1994, *Main Khiladi, Tu Anari* features Akshay Kumar, the macho star of Bollywood, as an upright cop who enlists the support of Saïïf, who plays the supporting role of a movie star, in his fight against criminals. In his review of this popular film, Row Kavi makes reference to one of the song-and-dance sequences in which Saïïf and Akshay are, as he puts it, "dancing holding each other, while their heroines are about ten feet away."[14]

In a recent essay, Thomas Waugh describes *Main Khiladi, Tu Anari* as a film that "has been widely recognized as the most vivid site of big things happening, both by the Mumbai gay circle around the magazine *Bombay Dost* and by local and foreign queer academics."[15] In a close reading of the now-infamous song-and-dance sequence that got both Row Kavi and Saïïf in trouble, Waugh writes:

> Akshay lifts Saïïf at the hips and carries him down the chorus line toward the camera; later the supine Saïïf, seen laterally, thrusts his groin upwardly toward the center of the frame, while Akshay, standing above him, is making similar pelvic thrusts toward the camera, so that the low angle perspective of the frame brings their thrusts together; toward the end of the number, the two dancers face each other and grasp each other's shoulders and Saïïf walks backwards vertically up a handy pillar, supported by Akshay, as the two maintain eye

contact. They literally can't keep their hands off each other: Akshay puts his tie on Saïïf, slaps his ass, even seems to touch his groin. But it is the larger, symmetrical bodily movements themselves that most play up the physical intimacy of the friendship bond, as they high kick in sync and prance together down the gauntlet of . . . girls.[16]

In the review published in *Bombay Dost,* Row Kavi also focuses his attention on the aforementioned song-and-dance sequence and, not unlike Waugh, concludes that *Main Khiladi, Tu Anari* "is a homosexual film."[17] According to Row Kavi, his review of *Main Khiladi, Tu Anari* in *Bombay Dost* was the background context that motivated Saïïf to beat him up, and the comments on *Nikki Tonight* "came as an added incentive." However, Saïïf and his family deny these allegations. In an interview with the *Times of India,* Saïïf's wife, Amrita Singh, who is also a Hindi film star, insists that "Ashok Row Kavi is a liar," because her husband "is incapable of raising his voice, let alone his hand, on a lady."[18] According to Singh, Row Kavi was "giving all sorts of statements to deviate from his major goof-up" on *Nikki Tonight.* "When he can say what he has about the Father of the Nation, can we expect a shred of decency from him?" she asks, adding, "His little tongue wags too much."[19]

Even if Row Kavi was attempting to "deviate attention" from "his major goof-up" on *Nikki Tonight,* he had caused plenty of controversy—too much for the incident to just go away. As fate would have it, Tushar Gandhi, a great-grandson of Mahatma Gandhi, was among those who chanced across the episode of *Nikki Tonight.* I also met with Tushar Gandhi in August 1996 to talk about this controversial episode.[20] Tushar Gandhi, by an interesting quirk of fate, lives just a couple of blocks down the street from Row Kavi in the Santa Cruz area of Mumbai. According to Tushar Gandhi, his encounter with the ill-fated episode of *Nikki Tonight* was "a very strange coincidence." He recalls his encounter with the show thus: "Usually, I used to switch [the TV] to the cable video channel which showed English films, and I would watch Hollywood movies. But on that day, at that particular time, they were showing some Hollywood film which was absolutely boring. So I started channel surfing, [and when] I switched to Star Plus, *Nikki Tonight* was on."[21]

The format of the show, as Tushar Gandhi recounts, was that the first guest came on, spoke for about a third of the show, then the next guest would come in and take about a third of the air time, and in the last third of the show, both guests would engage in free-for-all banter with the host, Nikki. On that particular day, Nikki's first guest was the popular Hindi film actress of yesteryear, Asha Putli, and Row Kavi was scheduled to follow her as the next guest. Says Tushar Gandhi:

It was a long time since I had seen her [Asha Putli]. She was an icon during my college time . . . you know, jazz and all that, and she was the original hippie. She started all those things, so, she was quite notorious. [Since] I was seeing her after a long gap of years, I said to myself, "Oh, there is Asha Putli, let's watch it." And then, of course, Ashok Row Kavi was introduced [as the second guest]. Immediately after his introduction, he uttered these words [calling Mahatma Gandhi a "bastard bania"].[22]

Tushar Gandhi recalls being "quite disturbed by the whole thing," and by the time the program finished, he had decided to act against it. And the rest is, of course, history. Outraged at the abusive reference to his great-grandfather and his family, Tushar Gandhi decided to take recourse to legal action. "It was like the proverbial last straw," says Tushar Gandhi. When people "shoot out their mouths" and abuse the name of Mahatma Gandhi, Tushar Gandhi finds that "except for the instantaneous furor," nothing else has been done to send out a signal that "this kind of thing will not be tolerated." At the same time, in the case of political leaders who have militant followers today, he argues, "no one even dares to do constructive criticism of those leaders because their followers are so intolerant that they don't stop at breaking bones."[23]

Tushar Gandhi claims that as a descendant of the Mahatma, "this was rankling him" for some time, and "anger was building up." He concedes that the *Nikki Tonight* episode was the first time he had taken up a case on the Mahatma's behalf. The reason, he admits, is that "unfortunately in the past, I was preoccupied with my own things, so I did not bother, except for feeling hurt and voicing my anger in private or giving a statement in the press. . . . But this time I decided that now there has to be a precedent. A signal has to be sent out that somebody will act on his behalf, and I did that."[24]

As he contemplated legal action, Tushar Gandhi says he considered two options: sue for defamation or start proceedings under the National Emblems' Act. Under this Act, any Indian citizen can appeal to the court of law when national emblems are defamed. As Tushar Gandhi recognized, under the Constitution of India, the name of Mahatma Gandhi—as the Father of the Nation—is protected by the National Emblems Act. He argues that if anyone desecrates the Indian national flag or other national emblems, then he or she is prosecuted. "Now, if you equate a person with national emblems, then an insult of his name should be treated as an insult of the nation," argues Tushar Gandhi.[25] However, if "the double standards against the Gandhi family," as he sees it, are to continue, then he demands that the portion of the law that equates Mahatma Gandhi to national emblems should be deleted. Explaining his position, Tushar Gandhi wrote to a member of Parliament from the region, Murli Deora, of the Congress Party, seeking an appropriate amendment to the Indian Constitution derecognizing Mahatma Gandhi's status as the Father of the Nation.

However, under the advice of his lawyer, M. P. Vashi, Tushar Gandhi did not purse this option further in the legal case against STAR TV after he realized that "the National Emblems Act leaves a lot of things unanswered."[26] For instance, even as the Act mandates that no one may use the national emblems for commercial benefit, it doesn't clearly specify what the legal consequences of such misuse would be. Given the ambiguity of the law in the context of the National Emblems Act, Tushar Gandhi decided to file the case in terms of defamation, which has a much more specific legal definition in the Indian Penal Code.[27] Under sections 499, 500, and 501 of the Indian Penal Code (IPC), defamation is defined as follows: "Whoever, by words either spoken or intended to be read, or by signs or by visible representations, makes or publishes any imputation concerning any person intending to harm, or knowing or having reasons to believe that such imputation will harm, the reputation of such person, is said, except in cases hereinafter excepted, to defame that person."[28]

The definition of defamation is further elaborated in the Indian Penal Code, with the following four explanations, the first of which, in Tushar Gandhi's view, is most particularly relevant to the *Nikki Tonight* episode.

> Explanation 1.—It may amount to defamation to impute anything to a deceased person, if the imputation would harm the reputation of that person if living, and is intended to be hurtful to the feelings of his family or other near relatives.
> Explanation 2.—It may amount to defamation to make an imputation concerning a company or an association or collection of persons as such.
> Explanation 3.—An imputation in the form of an alternative or expressed ironically, may amount to defamation.
> Explanation 4.—No imputation is said to harm a person's reputation unless that imputation directly or indirectly in the estimation of others, lowers the moral or intellectual character of that person, or lowers the character of that person in respect of his caste or of his calling or lowers the credit of that person or causes it to be believed that the body of that person is in a loathsome state, or in a state generally considered as disgraceful.[29]

Under the Indian Penal Code, the foregoing definition of defamation, in terms of the four explanations, holds as a rule, except for the ten following exceptions: (1) "imputation of truth which public good requires to be made or published"; (2) "public conduct of public servants"; (3) "conduct of any person touching any public question"; (4) "publication of reports of proceedings of courts"; (5) "merits of a case decided in Courts or conduct of witnesses and others concerned"; (6) "merits of public performance"; (7) "censure passed in good faith by person having lawful authority over another"; (8) "accusation preferred in good faith to authorized person"; (9) "imputation made in good faith by person for protection of his or other's interest"; (10) "caution intended for good of person to whom conveyed or for public good."[30]

As is evident from the foregoing list, while the explanations and many of the exceptions help add specificity to the general definition of defamation, some of the specified exceptions remain vague, and at best rather generalized. Therefore, in his filing of a "preliminary objection in law and prayer for stopping/dropping the proceedings against the accused," Row Kavi submitted to the Court that the Indian Penal Code does not clearly define the word "family" in the context of section 499. Row Kavi and his lawyer, Shrikant Bhat, argued that the defamation case in the *Nikki Tonight* episode was without merit because the meaning of the word "family" can only refer to people living together with the person who is defamed, or those who lived with a deceased person when he was alive. The word "family" cannot refer to a great-grandson or other descendants of the person defamed, for generations without limit, Row Kavi and his lawyer argued, because the acceptance of this meaning by the Court would extend criminal liability indefinitely into the future.[31] The Court, however, did not agree to the plea by Row Kavi to stop or drop the proceedings against him and cleared the way for Tushar Gandhi to pursue his case under the scope of defamation laws as specified in the Indian Penal Code.

Gandhi versus Murdoch

To initiate the court proceedings, Tushar Gandhi sent out legal notices to Row Kavi for uttering a defamatory remark, Nikki Bedi for hosting the show, Rupert Murdoch as the owner of STAR TV, Raghav Bhel of TV-18 as the producer, and Sanjay Roy Choudhari as the director of the show. The list of witnesses in the complaint filed by Tushar Gandhi included the names of President Shankar Dayal Sharma, Prime Minister P. V. Narasimha Rao, Home Minister S. B. Chavan, Information and Broadcasting Minister K. P. Singh

Deo, Maharashtra Chief Minister Manohar Joshi, the leader of the opposition parties in Lok Sabha, Atal Behari Vajpayee, the president of the BJP, L. K. Advani, the spokesman of the Janata Dal party, Jaipal Reddy, and the grandsons of the Mahatma, Arun Gandhi, Rajmohan Gandhi, and Ramchandra Gandhi.[32]

In the notice, Tushar Gandhi said that he would file criminal proceedings for defamation against the guilty parties and that he also planned to institute a civil suit claiming 50 crore rupees (approx. $12.5 million) as damages from STAR TV.[33] The notice further demanded that the show *Nikki Tonight* be stopped from being broadcast due to its vulgar and abusive language. Under the Indian Penal Code, a respondent can choose to prosecute the petitioner under a civil or a criminal litigation or both. As stated in the Indian Penal Code,

> the Civil Court would confine its decision to the trespass, threat of injury and damage by the servants, agents and workmen of the various defendants and the entitlement to token damages by the respondent, while the Criminal Court, the passage being per se defamatory, would proceed to find out whether any one of the 10 Exceptions to Sec. 499 IPC would apply. The scope of the two proceedings is entirely different, they are not parallel.[34]

Caught unawares by the civil and the criminal lawsuits, the network executives at STAR TV quickly apologized to Tushar Gandhi for the error, which they claimed had occurred due to carelessness on their part. In a press statement released on May 11, 1995, Gene Swinstead, STAR TV's managing director for India, Gary Davey, its CEO, *Nikki Tonight*'s Indian producers, TV-18, and the host of the show, Nikki Bedi, all issued an apology. In the statement, Swinstead was quoted as saying: "We fully appreciate that Mr. Ashok Row Kavi's comments caused deep offence, and while Mr. Kavi's comments and views are his own and do not in any way reflect those of STAR TV, Nikki Bedi, or TV-18, we at Star have a duty of care to hold the tastes and sensitivities of viewers paramount. I am therefore reviewing both the programme and our procedures with independent production houses."[35]

According to the press release issued by STAR TV, the decision to withdraw the talk show was made "in deference to and respect for the family of Mahatma Gandhi and the millions of its viewers to whom it may have caused offence."[36] Tushar Gandhi, however, responded that the legal proceedings would take their course irrespective of any apology that STAR TV had issued. He recalls that the day after the infamous incident, someone claiming to be from STAR TV called and spoke with him for over an hour. Putting the blame for the incident solely on the Indian producer of the show, TV-18, the STAR TV representative pleaded with Tushar Gandhi, "Don't be harsh on us. Go and do what you like to the producers—they are the bad ones."[37]

According to Tushar Gandhi, the STAR TV representative tried to convince him that "the Indian producers of the show had slipped the tape in, without permission from STAR TV, and thus the network was completely innocent."[38] But Tushar Gandhi remains unconvinced, and he counters STAR TV's claim to an innocent mistake with the following argument: "If tomorrow I send a smut film marked with the label of a popular STAR TV program, will they just air it? And if they run such a low ship, where they don't have any control over what they air, then they need to be penalized for that."[39]

Since *Nikki Tonight* was a prerecorded show, produced in India by TV-18, Tushar Gandhi argues that the producers must have carefully edited the Row Kavi interview before sending the ready-to-air tape off to STAR TV in Hong Kong. Posing a rhetorical question for the executives at STAR TV, Tushar Gandhi wonders whether, if Ashok Row Kavi had used the same words for political leaders who have very militant followers today, "Would TV-18 have still carried this comment, and would STAR TV have aired it?"[40]

I posed Tushar Gandhi's hypothetical question to representatives of STAR TV and TV-18 when I met with them in Mumbai in August 1996. However, both STAR TV and TV-18 refused to officially comment on any specific aspect of the case, with the justification that the matter was pending in the Court of the Chief Metropolitan Magistrate, at Esplanade in Mumbai. News reporters who contacted STAR TV in Mumbai for comment on the controversy shortly after it erupted had little success in eliciting an official response either. "It's being discussed at the highest management levels," was the only comment that STAR TV sources would provide.[41]

In the civil suit, Tushar Gandhi raised the damages amount from the initial sum of 50 crore rupees to 500 crore rupees (*crore* means "ten million") but later decided to drop the civil case entirely. When asked by journalists why he had sought to raise the damages amount and then had dropped the civil case, Tushar Gandhi replied: "That was my lawyer's decision. . . . I wasn't even too happy about the Rs. 50 crore, but he counseled me that in today's commercial world, this was the kind of action taken. Of course, I won't shirk responsibility for the decision; I was party to it, after all, but I have now decided not to claim monetary compensation."[42]

Tushar Gandhi claims that he was persuaded not to seek any monetary compensation when a journalist asked him how he had arrived at the price he was putting on his great-grandfather's reputation. "What is important is the principle," he argues, claiming that if the guilty are punished under criminal law, that is "good enough."[43] Under criminal law in the Indian Penal Code, the punishment for defamation is "simple imprisonment for a term which may extend to two years, or with fine, or with both."[44] However, unfortunately for Tushar Gandhi, the criminal case has not been proceeding as he had anticipated. One of the reasons for the lack of progress in the criminal case is that the process of issuing summonses was never successfully completed. Tushar Gandhi issued summonses to Row Kavi and TV-18 in Mumbai, but he has been unable to ensure the issuance of summonses to either Murdoch or Bedi, both of whom reside outside India.

Pleading his inability to summon Murdoch and Bedi, Tushar Gandhi requested that the magistrate issue a warrant for their arrest so as to ensure their presence in the Court. On July 3, 1995, the chief metropolitan magistrate, P. S. Patankar, issued a bailable warrant in the sum of 5,000 rupees against Rupert Murdoch. In his order, the magistrate noted that "there is prima facie case of insult and defamation."[45] The magistrate also issued summonses to Row Kavi, Bhel, and Choudhari, who live in Mumbai and therefore were able to appear in court. In a separate order, the magistrate also issued a search-and-seizure warrant instructing Mumbai police to search the office premises of STAR TV at 431 Lamington Road, Opera House, in Mumbai and confiscate videotapes of the controversial episode of *Nikki Tonight*.[46] Tushar Gandhi's lawyer, M. P. Vashi, however, expressed the fear that the cassette of the offending program might have been destroyed or gone missing.[47]

When I met with representatives of STAR TV, TV-18, Tushar Gandhi, and Row Kavi in August 1996, no one had a copy of the videotape. According to Row Kavi, his lawyer had requested the magistrate to ensure that Tushar Gandhi, as the complainant, provide them with a copy of the tape, which is a primary piece of evidence in the case. Tushar Gandhi, on his part, pleads helplessness in this regard. "There was no question of taping the episode," he argues, pointing out that he had accidentally chanced across the program when it was aired on STAR TV.[48] Therefore, Tushar Gandhi requested the magistrate to order TV-18, the producers of the show, to produce a copy of the episode in question.

If the loss of the primary piece of evidence wasn't enough to stall the progress of the legal proceedings, then what brought the case against STAR TV to a grinding halt was the fact that neither Murdoch nor any representatives of STAR TV based in Mumbai had come to face the trial. The Mumbai police reported back to the magistrate that the person named Murdoch was not located in the STAR TV office in the city, and thus indicated their inability to arrest him. On July 10, 1996, the chief metropolitan magistrate, M. L. Tahaliani, issued another bailable arrest warrant against Murdoch, this time, setting the bail amount at 1,000 rupees. While conceding that "in defamation cases the bail amounts are generally nominal," Vashi argued that the judge should have increased the bail amount to at least 2 lakhs, because "this is an issue which involves national prestige."[49] Tushar Gandhi joked that the bail amount in the arrest warrant was so low that "even Mr. Murdoch's peon can furnish it."[50]

Frustrated by the lack of progress in the case, Tushar Gandhi wrote to Prime Minister Gowda, seeking his help, but did not get any response. Instead, Prime Minister Gowda held a brief meeting with Murdoch on June 19, 1996, when the owner of STAR TV visited New Delhi to discuss plans for the expansion of his media empire in India. Since 1993, when Murdoch acquired the STAR TV network, he has been waging a protracted battle against the communist regime in China, which has denied his News Corporation legal access to a vast majority of the Chinese audiences.

Since the transfer of Hong Kong from Britain to China in 1997, a perceptible threat of stringent governmental regulations has loomed over STAR TV, which is the Asian base of Murdoch's global media empire. Recognizing the political significance of the changing media climate in Hong Kong, Murdoch began seeking out the more liberal governments of India and Taiwan, with the lure of shifting the already massive Asian base of his powerful media empire. However, even in the more liberal environs of India, where the government has allowed transnational media corporations like STAR TV legal access, Murdoch has been fighting a losing battle to acquire terrestrial broadcasting rights. Although there is now a thriving satellite and cable industry in India, the government does not grant terrestrial broadcasting rights in television to private companies, and longstanding laws have helped to preserve the monopoly of the state-run Doordarshan network in this regard.

Section 4 of the Indian Telegraph Act of 1885, which is one of the oldest laws in the country, gives the central (or federal) government the exclusive right to establish, maintain, and work telegraph networks across the country. This exclusive right of the Indian government in the transmission of telegraphy under the Telegraph Act of 1885 has been extended to other telecommunications services, including television broadcasting. Anyone who wants to legally transmit a television signal from within India has to necessarily seek a license from the Indian government for uplinking rights. Uplinking refers to the process of transmitting television signals from a ground station to an orbiting satellite, which then transmits the signals to viewership areas that fall under its transmission footprint.

To explore the possibility of gaining uplinking facilities within India, Murdoch met with Prime Minister Gowda on June 19, 1996, when Murdoch visited Delhi to discuss plans for the expansion of his media empire in India. Murdoch's visit, although a very low-key affair, quickly drew considerable media attention. It fueled rumors that Murdoch was keen to shift STAR TV's Asia base to India after the transfer of Hong Kong from Britain to China in 1997.

In a prominent, front-page story published in the *Business Standard,* government officials who spoke on condition of anonymity reported that Gowda had agreed to Murdoch's request "in principle," if STAR TV would abide by Indian laws and "uphold Indian values and culture."[51] According to a report in the *Economic Times,* Gowda is said to have assured Murdoch that the Indian government would be sympathetic to STAR TV's plans for setting up its Indian base in Tumkur, near Bangalore in Karnataka, which, incidentally, is the home state of the prime minister.[52]

Under this plan, STAR TV proposed to set up a state-of-the-art-studio, with postproduction facilities as well as facilities for uplinking, at an estimated cost of 5,000 crore rupees (approx. $145 million). Citing unnamed sources in the Karnataka state government, the *Economic Times* report suggests that the state's chief minister, J. H. Patel, took personal interest in this megaproject and handpicked a team of senior administrative officers to speed up the process for land acquisition and the setting up of telecommunications infrastructure to facilitate the smooth transition from Hong Kong to Tumkur.[53]

However, officials in the Indian government, the Karnataka state government, and STAR TV all vehemently denied the rumors about STAR TV moving to India. In Hong Kong, the STAR TV spokeswoman Jenny Poon scoffed at the reports, saying, "We have heard this [sic] kind of rumors in the past." Attempting to put all speculations to rest, she added, "Hong Kong will definitely be our base."[54] Meanwhile, Indian government officials described Murdoch's low-key visit as a "courtesy call." On its part, the Karnataka state government denied of any knowledge of the visit. Responding to questions about the meeting between Murdoch and Prime Minister Gowda, Chief Minister Patel commented: "I came to know about it only through the papers."[55] However, he added that if STAR TV's proposal was accepted at a later date, the state government would carefully study "its impact on the local media scene."[56]

Although Murdoch could have been apprehended by the Indian police under the arrest warrant issued against him by the Mumbai court, defamation does not constitute a criminal offense under which the extradition of an individual from another country can be sought. "Technically, Murdoch may not be fugitive from the law," agrees Tushar Gandhi, "but morally he is."[57] Claiming that he is not making "an irrational demand," Tushar Gandhi argues: "I am asking them to come to a trial which I can even lose." He points out that STAR TV has enough monetary resources to hire a battery of excellent legal brains. "Who is stopping them from doing that?" he asks.[58]

Therefore, Tushar Gandhi insists that his "main grouse" continues to be against STAR TV because, while they need India for their survival, "they are not willing to honor the judicial system" of the country.[59] However, in order to keep the legal proceedings in motion, Tushar Gandhi, under the advice of his lawyer, decided to separate the case against Row Kavi and TV-18—who have been issued summonses—from the case against Bedi and Murdoch, who are out of the country.

Row Kavi, on his part, protests against this attempt to separate the case against him from the one against Murdoch and STAR TV. "I am not the real target," he argues. "The gun has been kept on my shoulder, and it has been shot somewhere else. . . . It is so damn clear that they are after Rupert Murdoch."[60] To substantiate his claims, he refers to a conversation he had with the editor of an English newspaper, who, Row Kavi says, called him on the evening after the controversial episode of *Nikki Tonight* was aired on May 4, 1995. According to Row Kavi, this unnamed editor cautioned him that there had been "long conversations between three of the press magnates, who happened to be *marwadis,*" and that these media barons had planned "to make a big thing of this" controversy.[61] *Marwadi* is a term that is used to refer to a very influential business community in India, and some of the largest newspapers in the country are owned by businessmen who belong this community.

Row Kavi's recounting of a conversation with an unnamed editor is, of course, secondhand reporting of another conversation, and at best an unsubstantiated rumor. But his version of the story must necessarily be read in terms of the many reports making rounds in India that certain vested interests in the national media have supported Tushar Gandhi and encouraged him to pursue the case against Murdoch. Another version of this rumor is that powerful business interests in India have used the *Nikki Tonight* controversy to mobilize public opinion against Murdoch to ensure that trans-national media corporations like News Corp are kept out of the domestic print media industry, which has been protected by government regulations from direct foreign competition.[62] Responding to the rumor that he is being "used as a weapon" by certain vested interests in the Indian media, Tushar Gandhi concedes that it may have happened, but he argues: "I don't go out of my way to be exploited that way. . . . Many people have accused me of being a front-man for the *marwadi* press barons, as they are called. But my conscience is clear. I have not taken any help, except for the media publicity that I get in the process. I have not taken any help, so why bother [about these speculations]?"[63]

Nonetheless, Tushar Gandhi's desire to use the media publicity from the controversy surrounding his campaign against the abuse of his great-grand-father's name seems to only add fuel to speculations, suspicions, and con-spiracy theories. On October 2, 1995, Tushar Gandhi decided to undertake a *satyagraha* (civil disobedience) protest to stop politicians from going to the Mahatma's memorial at Rajghat in Delhi to express his displeasure at "how the government was dragging its feet on the STAR TV case."[64] He also ex-pressed his commitment to conducting other peaceful protests if the court case against STAR TV did not progress satisfactorily.

Soon after the publicity surrounding the *Nikki Tonight* controversy sub-sided, Tushar Gandhi was again in the news when he scattered the last remain-ing ashes of his great-grandfather at the holy site signifying the confluence of the Ganga and Yamuna rivers in the north Indian city of Allahabad. On Jan-uary 30, 1997—the forty-ninth anniversary of the Mahatma's assassination—in full view of television cameras from around the world, Tushar Gandhi emptied the ashes from a copper urn into the river, as prescribed by the tradi-tions of a religious ritual that many Hindus perform as the last rites of a de-parted relative. The copper urn containing the remnants of the Mahatma's ashes had been lying forgotten for decades in a vault of the State Bank of India in the eastern Indian city of Cuttack. When the urn was accidentally discovered in 1995, Tushar Gandhi filed a case in the Indian Supreme Court and won the right to possession of the ashes.

After collecting the urn from the bank vault in Cuttack, Tushar Gandhi made a much-publicized 600-mile journey to Allahabad in a special railway train. The urn was displayed for one week to allow members of the public to pay their last respects to the Father of the Nation, before his ashes were dispersed in the holy rivers for one last time. According to news reports, hundreds of people filed past the urn to pay homage, but the ash-scattering ceremony had a much lower turnout than the state government officials overseeing the event had anticipated. Prime Minister Gowda, who had indicated his interest in attending the ceremony, was conspicuous by his absence. Dismissing this public snub from the highest political official in the government of India, Tushar Gandhi declared: "I would be gratified even if one person came to pay homage to Mahatma Gandhi sincerely."[65]

On August 15, 1997—the fiftieth anniversary of India's independence from British colonialism—Tushar Gandhi felt snubbed by the national government again, when no one from the Gandhi family was invited to a state-sponsored ceremony remembering the Mahatma's leadership in the freedom struggle. Calling it "a shocking lapse" by the government of India, Tushar Gandhi expressed his disappointment at how the "descendants of the Father of the Nation have been virtually forgotten even as all political parties exploit Gandhi's name to boost their image."[66]

Tushar Gandhi argues that politicians of various parties—be it the Congress Party, the BJP, or the Janata Dal—have turned the Mahatma into a cheap publicity vehicle and seek to cash in on the Gandhian charisma without following any of his principles. Claiming that India needed "good people" to begin a "cleansing process" in politics, Tushar Gandhi jumped into the electoral fray during the general elections of 1998 by taking on the right-wing Shiv Sena, which controls the local government in Mumbai and shares the political philosophy of Hindutva with the BJP at the national level. Although he lost the elections to the Shiv Sena candidate, Tushar Gandhi's campaign to fight the "unscrupulous forces of fascism" seemed to resonate with two important constituencies: (1) wealthy businessmen and film tycoons who were the targets of extortion by underworld dons with connections to powerful leaders in the Shiv Sena, and (2) Muslim minorities and lower caste communities who were the victims of communal violence and organized crime in the city of Mumbai. Asked by reporters how he felt about allying with the socialist Samajwadi Party—some of whose leaders have been extremely critical of Mahatma Gandhi—Tushar Gandhi replied: "I am not [one] hundred percent Gandhian."

Row Kavi finds it intriguing that Tushar Gandhi's interest in protecting the sanctity of the Gandhian mantle was only aroused after the *Nikki Tonight* controversy erupted in 1995. To support his argument that "something very mysterious is going on," Row Kavi poses a series of rhetorical questions:

> How is it that he [Tushar Gandhi] is always complaining about money problems, but he has got money to fight an expensive case like this? . . . How is it

that he has not protested about Mrinal Sen [the noted Indian film director] saying in Jadhavpur University during the Convocation ceremony that [Mahatma] Gandhi was born from the anus of a pig? ... How is it that Tarun Gandhi [*sic*] did not say a word [when] Mayawati [of the Bahujan Samajwadi Party] along with a group [of her followers] threw five to ten buckets of shit on Raj Ghat [Mahatma Gandhi's memorial in Delhi]? How is it that this great Gandhi family then didn't feel it?[67]

In posing these rhetorical questions, Row Kavi argues that what has become of Gandhism today is "the biggest farce." In Row Kavi's view, Tushar Gandhi should protest more about the Mahatma's family being silent witness to these perversions of Gandhism by powerful political leaders "and not at what is being said about Gandhi" by marginal individuals like him. Mayawati's critique of Gandhian nationalism for its marginalization of the lower castes, according to Row Kavi, is as legitimate as is his criticism of Gandhi's own inability to come to terms with issues of homosexuality in the national community.

"All down the line," argues Row Kavi, "Gandhism has betrayed what could have been modern India."[68] His frustration about the marginalization of the gay community in the national mainstream, Row Kavi argues, was the backdrop for his controversial letter written in 1971 to make apparent the many contradictions of Gandhian nationalism in India. Acknowledging that he made a mistake on *Nikki Tonight*, Row Kavi says: "I am not saying that what I have done is right. When it comes out in court, I will explain that. ... The man [Mahatma Gandhi] really was someone great, and Einstein wasn't wrong [in saying that generations to come will scarcely believe that such a man in flesh and blood walked among us]. But please don't make him out to be a god."[69]

Perhaps not many in the 1970s had taken notice of the "bitterness" of the marginalized gay community against the many contradictions of Gandhian nationalism in modern India that Row Kavi had sought to express through his controversial letter to the editor of the *Illustrated Weekly of India*. Even if his letter to the *Weekly* was deliberately mischievous, one must acknowledge that as one of India's leading and most prominent gay spokespersons, Row Kavi has constantly faced discrimination in a society where it is extremely difficult for gays and lesbians to come out. Under section 377 of the Indian Penal Code, homosexual activity is defined as a crime punishable by life imprisonment. In 1983, the Supreme Court upheld the conviction by a lower court of a man who had engaged in a homosexual act, reducing the sentence from three years of rigorous imprisonment to six months because the two men had consented to sexual intercourse.[70]

Although there is a tradition of tolerating sexual differences in Hinduism, Row Kavi argues that in contemporary India, "there is an unwritten law that if you are not married you don't exist in society."[71] In his own family, Row Kavi explains, homosexuality is never a subject of discussion, and invitations to social functions, which are usually sent to the eldest son in the family after the father's death, are sent to his mother instead. Therefore, Row Kavi has taken it upon himself to fight against the marginalization of the gay community by bringing to light what he calls "the deep homosexual and lesbian roots in India's history and religious practices."[72]

Row Kavi refers to the existence in India for centuries of a special group called "hijras" ("not men"), who are invited to dance at weddings and other important ceremonies because it is considered "auspicious" to have their blessings. He also points out that the Hindu god Skanda, the son of Shiva and Vishnu, is considered the patron saint of gay men, and millions of pilgrims visit his shrine every year. However, Row Kavi is concerned that "there are only loose gay networks across Asia," and there is no political, legal, or social support for establishing a gay community or identity in India.[73]

In spite of, or perhaps because of, the suspicion, hostility, fear, ridicule, and uneasy tolerance that homosexuality encounters in Indian society, Row Kavi has always put himself on the line by being outspoken, and oftentimes outrageous, in his criticism of the nationalist mainstream. In November 1994, the novelist and gossip columnist Shobha De wrote a profile of Row Kavi for *Week* magazine in India, describing him as an "agent provocateur" who constantly "plays up to all the cliches associated with overt gay conduct . . . deliberately and mischievously . . . to see the horrified expression on people's faces."[74] (See figure 11.)

De, who claims years of personal friendship with Row Kavi, remarks that there is also the "other" Row Kavi, "and to me, the real person," who is "a very different man" and has taken on the establishment "at extraordinary risk to himself," to render "invaluable service to countless anonymous gays in India and outside." De reminds us that it was Row Kavi who was the moving spirit behind the magazine *Bombay Dost*.

The puritan condemnation and uproar that followed the publication of the first Indian magazine catering to gay and lesbian communities made Row Kavi "the target of a systematic hate campaign," and "his phone lines were jammed with dirty calls and nasty threats."[75] De also recounts how in

Figure 11. Caricature of Ashok Row Kavi in the
Week, November 6, 1994, 19. Courtesy, the *Week*, a
Malayala Manorama publication.

the early years of HIV hysteria and the prejudice against people with AIDS,
Row Kavi was the force behind the organized effort to demystify the disease
in the bustling city of Mumbai. Gradually, Row Kavi's contribution to the
fight against AIDS was discovered by the mainstream media, and with it
"came a grudging sense of respect." De writes: "The man who was once
labelled a 'pervert' and 'degenerate' . . . was now seen as the new messiah—
the savior of a deeply misunderstood minority community."[76]

Perhaps the curious reputation as a provocative loudmouth and a misunderstood missionary of the gay community that Row Kavi has earned in
the influential media circles of Mumbai was what induced the producers of
Nikki Tonight to invite him on the talk show, which had cultivated a similar
reputation of being outrageous and outspoken. In a country where the talk
show genre is a relatively new phenomenon, *Nikki Tonight,* which made its

debut on Star Plus in March 1995, was something of a curiosity, confusion, and fascination for television audiences in India. In fact, *Asian Advertising and Marketing* reports that a month before the Row Kavi episode put an end to the show, STAR TV "had been touting *Nikki Tonight* as one of the best and most exciting talk shows to be aired on Star Plus in India."[77]

In the Age of Glocalization

Since May 1991, when STAR TV began broadcasting over Asia, using a satellite from Hong Kong, cable television viewers in India found themselves exposed to a variety of hitherto unheard voices and unspoken messages on television. Talk shows on STAR TV, like *Springer* and *Ricki*, may well seem to Indian audiences to be a shocking exhibition of marginalized individuals and fringe communities in American society. Hindi talk shows on Zee TV, like *Shakti* and *Teer Kaman*, which handle controversial issues about women and children in Indian society, may well have earned the mantle of Hindi *Oprahs* in India. English talk shows, like *Chat Show*, on the state-owned television network *Doordarshan*, may have given Indian viewers a taste of the Leno/Letterman model of urbane flirtations with celebrities. But there was something very different about *Nikki Tonight*.

First, unlike those on *Springer* or *Ricki*, the guests on *Nikki* were all Indian. Second, in contrast to the Hindi talk shows modeled on *Springer* or *Ricki*, the guests on *Nikki* were no ordinary folk who could be dismissed as the societal riffraff. The guests on *Nikki Tonight* were all recognizable Indian celebrities who were well known, at least among the Anglophile viewers that the talk show attracted. However, distinct from the Indian *Chat Show*, in the Leno/Letterman model of urbane socialization, *Nikki Tonight* veered toward outrageous vulgarity. Nikki's flirtatious antics with celebrities, mischievous comments about their personal lives, and cheeky questions about their sex lives appeared seductively enticing, even as they seemed to go way beyond conventional norms of social behavior in Indian society.

For instance, on the very first episode, when Nikki unabashedly discussed the size and the feel of her silicone-implant breasts with her film star husband Kabir Bedi (who was her first guest), it shocked the modesty of viewers not used to hearing women, certainly not Indian women, publicly detail their anatomy. Or when Nikki impudently asked the eighty-year-old doyen of Indian tabloid journalism, Russi Karanjia, questions about his virility since

recovering from a bad bout of flu, it seemed outrageously insolent in a society where one doesn't usually speak of sexuality and the elderly in the same
breath.[78]

But Nikki did not find it offensive in the least. "We are only interested in
having a friendly, cheeky conversation with our guests, not in tearing them to
bits," she claimed in an interview given a few weeks after the launch of the
show.[79] In another interview, Nikki declared: "I have to make the show racy
and unpredictable to keep the viewer glued, so I avoid the usual run of the mill
questions."[80] In those early, heady days of success for *Nikki Tonight,* journalists
interviewing Nikki saw her as a "a confident media-savvy woman picking up
compliments like 'sexy, stylish and beautiful,' which she may not have been
showered with a few years ago."[81] On the other hand, there were the critics
who saw in Nikki only a "dazzling smile," which, they argued, "concealed an
absence of much brain activity."[82] Not surprisingly, then, Nikki herself was
somewhat of a novelty, curiosity, and even aberration for Indian television
viewers (see figure 12).

The daughter of an Indian father and an English mother, the blonde-haired
Nikki was raised in England and moved to Mumbai in the late 1980s in search
of a film career. After a few roles in popular Hindi TV serials on Doordarshan,
such as *Mr. Yogi, Kab Tak Pukaroon,* and *The Sword of Tipu Sultan,* Nikki starred
as Desdemona in Alyque Padamsee's *Othello* in Mumbai theater. Here, on the
sets of *Othello,* Nikki met and married the man playing Iago—the transnational
Indian film star Kabir Bedi, who is almost twice her age.

The romance between the older, much-married Kabir Bedi and young
Nikki, who was then married to the food stylist Sunil Vijakar, was quite a
scandal, even for the tinsel town of Mumbai, which is no stranger to such
controversy. Kabir Bedi, who has acted in Hollywood films like *Octopussy*
and appeared on television soap operas like *The Bold and the Beautiful,* has
made a career for himself in the United States. Nikki, who has to divide her
time between her husband in Hollywood, her family in England, and her
own career in India, describes herself as a "glocal girl"—"global plus local"—
ideally suited to host a "glocal" show about "glocal" Indians.[83] So in July
1994, when TV-18 contacted her to audition for a planned twenty-eight-
episode talk show on STAR TV, Nikki "jumped at the opportunity."[84] Although "the competition was stiff," Nikki's "pizzaz [*sic*] and hybrid pretti-
ness . . . helped." Also of help was her experience hosting, for Channel Four
in Britain, a show called *Bombay Chat,* which, Nikki reveals, "was mainly
made to acquaint second-generation Indians in Britain with the land of

Nikki herself was too busy being herself and did not remonstrate

Figure 12. Caricature of Nikki Bedi by R. K. Laxman
in the *Times of India*, May 14, 1995, 10.

their immigrant parents."[85] Nikki's creative neologism, defining her hybrid
identity as a "glocal girl," aptly captures the many contradictory dimensions
of what Arjun Appadurai has outlined as "disjuncture and difference in the
global cultural economy."[86] For Appadurai, the disjunctive cultural flows
that at once sustain and disrupt the global economy are

[1] ethnoscapes produced by flows of people: tourists, immigrants, refugees,
exiles and guest workers. . . . [2] technoscapes, the machinery and plant flows
produced by multinational and national corporations and government agencies. . . . [3] finanscapes, produced by the rapid flows of money in the cur-
rency markets and stock exchanges. . . . [4] mediascapes, the repertoire of
images of information, the flows which are produced and distributed by

newspapers, magazines and film. . . . [5] ideoscapes, linked to flows of images which are associated with state or counter-state movement ideologies which are comprised of elements of freedom, welfare, rights, etc.[87]

Appadurai's theorization of the "scapes" in globalization help us to better understand some of the reasons for the rise of hybrid "glocal" shows like *Nikki Tonight* on Indian television in the 1990s. If the notion of mediascapes, for instance, can help us to understand how media networks like STAR TV can at once claim to be multinational and multilocal, then the notion of technoscapes can explain how these "glocal" networks use satellite and cable technologies to at once broadcast globally and narrowcast for local audience preferences. Similarly, Nikki's self-definition as a hybrid "glocal girl" signifies how the disjunctive flows of ethnoscapes have disrupted traditional organization of "identity" and "community" in terms of national geographies and cultural histories in the dynamic flows of electronic capitalism.

However, the controversy engulfing television shows like *Nikki Tonight* also reveals that, even as we are integrated in an increasingly globalized or, more aptly, glocalized world of satellite television, the traditional issue of national identities and cultural boundaries has become ever more important. The symbolic significance of this mechanism is evident when a transnational media network like STAR TV transgresses national boundaries to entertain local viewers with "glocal" stars like Nikki Bedi who proclaim: "we are here to have fun and not hammer people on their beliefs or politics."[88] Such rhetoric grossly underestimates the potency of nationalist imaginations that draw upon the community's collective sense of familial and cultural heritage. Stuart Hall makes this point elegantly in his discussion of British identity politics in relation to what he calls "the very contradictory space" of the "global postmodern."[89] Hall argues:

> We have just, in Britain, opened up the new satellite TV called "Sky Channel," owned by Rupert Murdoch. It sits just above the Channel. It speaks across to all the European societies at once and as it went up all the older models of communication in our society were being dismantled. The notion of the British Broadcasting Corporation, of a public service interest, is rendered anachronistic in a moment. . . . At the same time as sending the satellite aloft, Thatcherism sends someone to watch the satellite. So Mrs. Thatcher has put into orbit Rupert Murdoch and the "Sky Channel" but also a new Broadcasting Standards Committee to make sure that the satellite does not immediately communicate soft pornography to all of us after 11 o'clock when the children are in bed.[90]

The "balancing act," as Hall puts it, of Thatcherism, which attempts to regulate "the already deregulated Rupert Murdoch" with "an old petite bourgeois morality," in Britain is clearly at play in Indian nationalism, as it seeks to mediate diverse imaginations of collective identity and cultural differences in the dynamic flows of electronic capitalism. For instance, when Nikki prompted Row Kavi to recall his abusive remarks against Mahatma Gandhi, neither the Indian producers of *Nikki Tonight* nor the media executives at STAR TV had perhaps imagined that the old "bourgeois morality" of nationalist consciousness would be so powerfully mobilized to protest a stray transgression on a flippant talk show. Nor had they considered the possibility that someone from Mahatma Gandhi's family would take issue with Rupert Murdoch and his pan-Asian television network and start a systematic political and legal campaign against them. What drew the Gandhi family's ire, according to Tushar Gandhi, is that STAR TV could easily have edited the offensive reference, and it wouldn't have affected the interview in any way. Elaborating on why STAR TV did not do so, Tushar Gandhi argues:

> The program thrived on notoriety, it had a history in its short run of doing sensationalistic things, asking sensationalist questions to interviewees, and encouraging them to give sensational, outrageous comments. . . . They must have felt that by airing this comment [by Row Kavi], they would create an uproar, which would die down after a measly little apology, and the viewership of the program would go up tenfold. And thus, they would benefit from the whole thing.[91]

Moreover, says Tushar Gandhi, STAR TV did not realize that he would carry out a sustained campaign against them. He argues that the controversy "created a tremendous furor" and even reached the Indian Parliament because he "didn't let the thing die down."[92] Immediately after the controversial incident, Tushar Gandhi, through his lawyer, wrote to the President of India and the leaders of various political parties. He requested these leaders to "vindicate the honor of not only Mahatma Gandhi but Indians in general and the Gandhi family in particular."[93]

Responding to Tushar Gandhi's call, furious politicians, in an exceptional show of unity across party affiliations, referred to Row Kavi's derogatory remark against Mahatma Gandhi in the Indian Parliament (the remark was promptly expunged by the chair as a mark of respect for the Father of the Nation). Raising the issue in the Rajya Sabha, the upper house of Parliament, Mohamed Salim, a member of the Communist Party of India-Marxist (CPI-M), said that "it was a matter of shame the Father of the Nation had been maligned." He admonished the government for preferring to remain silent "when imperialist forces were out in the open."[94]

Supporting his Marxist colleague, K. R. Malkani, a member of the right-wing BJP, remarked that the scandalous talk show amounted to "cultural terrorism" and demanded that the government ask STAR TV to "shut shop" in India. Jaipal Reddy, of the socialist Janata Dal party, argued that *Nikki Tonight* was a clear act of vandalization of India's heritage. "We are outraged," he announced in Parliament. "We are all inheritors of Gandhi's legacy and we should take it as an insult to the nation."[95]

As agitated members across party lines rose to defend Gandhi's legacy, the then union home minister, S. B. Chavan, quickly agreed that "there was no scope for a difference of opinion on the issue."[96] He assured the members that the government would have the legal department carefully examine the issue. "If the legal department won't come to our rescue," he added with uncharacteristic bravado, "then we might have to go in for a new legislation" to ban the transborder transgressions of media networks like STAR TV.[97]

In the unanimous agreement that there could be no difference of opinion about the unspeakable remark against the Father of the Nation, the Indian Parliament was attempting to demonstrate how ideological differences must be transcended to sustain what Benedict Anderson has called the imagined community of a nation.[98] However, as the controversial episode of *Nikki Tonight* clearly reveals, irreducible differences within the national community were already in evidence, even as the Indian Parliament was seeking to sustain the collective imagination of the nation in the name of its venerated patriarch. First, Nikki's hybrid identity as a "glocal girl" can hardly be considered the symbol of imperialist forces out to get Indian nationalism. Second, Row Kavi is not a "cultural terrorist" of a hostile country but an Indian citizen venting his frustration at the marginalization of the gay community in the national mainstream. Citing Lawrence Cohen, who has written extensively on issues of homosexuality and gender relations in India,[99] Row Kavi accurately captures the irony of the political protests over the *Nikki Tonight* controversy in the nationalist mainstream thus: "One is a blonde woman who claims to be an Indian, and [the other] is someone who is openly homosexual. The elite of India sees both these constructs as threatening. Both are penetrative concepts which upset them."[100]

Given the onerous responsibilities of holding the nation together against the transgressive potential of penetrative concepts of gendered hierarchies and sexual differences that could threaten its collective imagination from within and without its boundaries, should one commend the leftist and rightist parties for momentarily subsuming ideological differences and rushing to defend the Name of the Father of the Nation? Especially so, given that the Congress Party—which claims the mantle of the proud Gandhian legacy in Indian politics—maintained a "guilty silence," in its refusal to take a stance against STAR TV until new legislation was enacted to ban the transmission of transnational networks. However, even as a number of leaders from the left and the right of the political spectrum criticized the centrist Congress Party for its inability to defend Gandhi's reputation as the Father of the Nation, the inescapable reality was that they were speaking for a man whose pacifist politics has always been at odds with their professed ideologies of revolutionary Marxism and Hindutva nationalism, respectively.

Moreover, if the Congress Party sought to deflect the criticism of its "guilty silence" by soliciting support in Parliament to legislate a ban on satellite networks like STAR TV in India, then the other national parties also came woefully short of mandating a law to punish those guilty of tarnishing the name of the Father of the Nation that they all claim to so revere. As already mentioned, when the Congress Party lost the general elections of 1996, a coalition of socialist, communist, and regional parties came to power in a coalition government led by Prime Minister Deve Gowda of the socialist Janata Dal. In the same year, when Murdoch visited India to explore the possibility of locating the base of his Asian operations in India, Prime Minister Gowda met with him in New Delhi to discuss the details of the proposal. Although Gowda and officials in the Indian government were aware of the arrest warrant issued by the Mumbai court against Murdoch in the *Nikki Tonight* case, clearly they were willing to ignore the law of the land that they had sworn to uphold, so as to avoid any political embarrassment. A year later, following the fall of the left-wing Gowda government, a new government led by the right-wing BJP came to power. But, the BJP-led government—some of whose leaders openly embrace a xenophobic ideology of Hindutva nationalism—did little to arrest the growth of STAR TV in India (not to mention the pending arrest warrant against its powerful owner, Rupert Murdoch).

Is There an Indian Community of Television?

If the nation is, as Anderson argues, an imagined community sustained by a collective sense of national identity, then the Indian Parliament—representing the collective imagination of the nation—it appears, miserably failed the nationalism test, even in the face of a transient transgression on a flippant talk show. A stray remark on television against its most exalted symbol—the Father of the Nation—revealed the woeful inadequacies of imagining the nation as a collective entity in the face of the irreducible differences of transnational and translocal networks from within and beyond its borders.

However, despite the failure of the Indian Parliament to sustain the collective imagination of the nation, one cannot, and must not, hastily announce the death of nationalism in the age of electronic capitalism. For, ironically enough, the inability of the Indian Parliament to assert the collective will of the nation was highlighted precisely at the moment when political parties across the spectrum "came together" in their criticism of the transborder transgressions of STAR TV. That is, at the very moment when the empirically extant community of Indian nationalism was disrupted by a momentary transgression on a transnational network, it also avowed its collective sense of identity.

The question of an Indian community of *television,* I conclude, cannot be adequately answered by merely asserting national identity, or celebrating cultural differences. Instead, I posit that the interplay between national identity and cultural differences in Indian television can only be addressed in terms of what I call *unimaginable communities,* which, however paradoxical it may appear, are at once based on identity/difference.

As I define them, unimaginable communities are empirically extant, extinct, or imagined communities that fleetingly avow their collective identity in the face of irreducible differences. However, it is the paradoxical nature of the unimaginable community that at the very moment that it avows its identity, it loses all claims to legitimacy, and gives voice to irreducible differences. By invoking the paradoxical notion of the unimaginable community in the *Nikki Tonight* controversy, it becomes possible to explain how and why there is a renewed resurgence of nationalist imagination in Indian television at the very moment when marginalized individuals within India assert their cultural differences.

The notion of the unimaginable community also helps explain how and why at the very moment when the national community of Indian television avows its collective identity, it loses its claim to legitimacy, and gives voice to the irreducible differences of transnationalism and translocalism within and beyond its borders. One may argue that the notion of the unimaginable community is an obscurantist reformulation of the Freudian return of the repressed or the Marxian revolt of the oppressed. However, I would argue that both Freudian and Marxian notions of community overindulge in imaginations of nationalist identity that never is. On the other hand, to speak of the nation as an unimaginable community is to speak of an unspeakable transgression of an unspoken identity that never will be. To speak of the unimaginable community is, in this sense, to speak of the eternal return, or the unceasing presence, of the transgressed.

But how does one speak of the eternal return of transgressions? How can one speak of—or speak for—the unspoken identity of an unimaginable community? Georges Bataille provides subtle insights into the process of imagining the unimaginable in the following terms: to imagine a community into existence is to strive toward the knowledge about that which one presumes to not know or have no knowledge of.[101] In other words, the work of imagination is predicated on a paradoxical moment where the acquisition of knowledge about one's sense of community occurs through a process of loss that I describe as "unimagination." The notion of unimagination—as a momentary sense of loss—is of particular relevance to my critique of postcolonial nationalism, where the ideal of community is founded in terms of a lack of imagination. For instance, secular imaginations of Indian nationalism, as described in Nehru's *Discovery of India,* are founded upon a sense of identity that the community is deemed to be lacking in a most acute way (due to the pervasive diversities of religion, region, ethnicity, language, class, caste, gender, and sexual orientation).

Similarly, Hindu fundamentalist ideals of Indian nationalism are based on an essentialist notion of Hindutva (pure Hindu identity) that, according to its proponents, the community is lacking in the most fundamental way (due to the influx of foreigners and minorities, who are seen as corrupt influences on the nation's original essence). Therefore, in both its secularist and fundamentalist incarnations, the "national family" is imagined in terms of a sense of lack that the community must collectively overcome in order to attain the idealized state (of secularism and religious nationalism, respectively).

In these ideological contexts of postcolonial nationalism, the community is required—often through state-sponsored laws—to collectively embrace certain symbols and images that are considered emblematic of the values that the nation is imagined to be lacking as a whole. Therefore, the most cherished symbols of nationalism (such as the flag) and the most exalted icons of the national family (such as the name of the Father of the Nation), or the most ancient myths of the national culture (Hindu epics such as the *Ramayana* and *Mahabharata*), are considered very precious in the

state-sponsored project of imaging a sense of identity that the community
is deemed to be lacking as a collective.

Therefore, excessive criticism of the nationalist ideals that bring into
question the exalted status of its icons—such Row Kavi's insulting remarks
about the Father of the Nation on STAR TV—is seen as an unpardonable
transgression. Since the *Nikki Tonight* controversy on STAR TV, there have
been many other cases in India in which electronic media such as television
and other mass media such as cinema and theater have been criticized, regulated, or sometimes even banned in the nationalist mainstream when considered to be a threat to the collective imagination of national identity that
the community is presumed to lack.

In July 1998, the state government of Maharashtra was pressured to ban
the play *Mee Nathuram Boltoy* (I Am Nathuram Speaking), as critics mount-
ed a national protest against its portrayal of Mahatma Gandhi's assassin,
Nathuram Godse, as a heroic figure. In the play, Nathuram declares "Gandhi
vadh" as a national duty. "Vadh" in Marathi (and in Hindi) means murder
or assassination, and also carries a religious connotation of the victory of
good over evil in Hindu mythology. The play also has a comment by Godse
that Gandhi didn't die with the words "He Ram" (O God) on his lips, and
that he only murmured "Huh" as he fell to the ground. That the director of
the play, Vinay Apte, is affiliated with the BJP only complicates matters, since
many of the party's critics characterize its Hindutva brand of nationalism
as the moving force behind Godse's decision to kill Gandhi.

Leading members of the Congress Party, such as Pranab Mukherjee and
H. R. Bharadwaj, raised the issue of the controversial play in Parliament and
forced the presiding officer to adjourn the day's proceedings until a satisfac-
tory resolution was reached. The home minister, L. K. Advani, assured the
agitated Congress Party members in Parliament that he had advised the Maharashtra government to ban the play, and expressed confidence that his advice
would be heeded by the state officials. Leaders of the Congress Party in the
state led the charge for the government ban and disrupted the staging of the
play at the Shivaji Mandir Theater in central Mumbai.[102] However, news reports
also indicate that the venues where the play was shown in Mumbai were sold
out, and additional requests for showings came pouring in from cities and
towns around the state.[103]

In another play, *Gandhi Virudh Gandhi,* Gandhi is portrayed as not just
a deified symbol but a complex historical figure who is "a Bapu who failed
as a *baap,* a Mahatma who could convince a nation but could not communicate with his son, a leader who failed as a father."[104] The noted film, televi-

sion, and theater director Feroz Khan, who translated the play into English, as *Mahatma vs. Gandhi,* acknowledges the enormity of the challenge of depicting the "private pain of a public individual." Naseeruddin Shah, who plays Gandhi in the English version, describes the play as "the story of a troubled father who happens to be the 'Father of the Nation.'"[105] Although a play such as *Mahatma vs. Gandhi* endeavors to portray Gandhi as a human being with many failings, Anagha Sawant argues that "strangely enough this has not defiled the Mahatma."[106] Instead, Sawant writes,

> today, Mohandas Karamchand Gandhi comes across as more human than ever before. Bapu has become a more tangible construct, on one hand, and a less sensitive commodity on the other. Now, be it the Mahatma's relationship with his first born, Harilal (Gandhi Virudh . . .), the friction of his line of thinking with that of Baba Saheb Ambedkar or his understanding of the Father of the Constitution (Gandhi-Ambedkar)—the Mahatma is being treated like a man. Last weekend, the final boundary was crossed with Mi Nathuram . . . when theater looked at Gandhi through the eyes of his assailant, Nathuram Godse.[107]

The controversy and the success of plays like Vinay Apte's *Mee Nathuram Boltoy,* Chandrakant Kulkarni's *Gandhi Virudh Gandhi,* and Centan Datar's *Gandhi-Ambedkar*—and one could add the *Nikki Tonight* controversy to this list—are good indicators of the growing debate in India about Gandhi's significance, not only in terms of his symbolic role of the Father of the Nation but also in relation to his historical place as a father and political leader with many lofty ideals and many more failings. In this context, irreverence toward nationalist icons like Mahatma Gandhi on television shows, such as *Nikki Tonight,* is particularly problematic for political elites at the helm of state affairs, since the pervasive power of electronic media, it is feared, can dramatically hinder the imagined sense of identity in the national community. One has only to flip through the channels in India to see how talk shows, sitcoms, news shows, or music videos often ridicule and criticize the state-sponsored agenda of protecting iconic figures such as the Father of the Nation—much to the consternation of the political, economic, and cultural elites who see such representations on television as unmitigated threats to the very survival of the nation as a community.

However, in positing the necessity of cultural transgressions on Indian television to aid the deconstruction of nationalist imaginations, I seek to invoke a more archaic meaning of community, one that is articulated not in terms of lack but through a process of loss; that is to say, through the expenditure of excess. Of particular significance to this discussion is the theory of gift-exchange propounded by Marcel Mauss in 1925, in his well-known essay *The Gift*. The central function of gift-exchange in archaic communities is to foster a process of communication and sense of communion among their members. The person who receives a gift is obligated to return the favor, perhaps even better it. In such a community, the worth or the status of an individual is determined not by how much wealth one accumulates but by how much one gifts, or donates (*don* in French, *dana* in Sanskrit). In a community based on gift-exchange, one who aims to accumulate wealth and never expends excess becomes a social outcast.[108]

The general status of excess in a community based on the principles of gift-exchange, Jean Piel argues, "throws light on a large number of social, political, economic and aesthetic phenomena," since "luxury, games, spectacles, forms of worship, sexual activity (set apart from the finality of genital function), the arts and poetry in the strict sense of the term are together so many manifestations of improductive expenditure."[109]

To the list of cultural phenomena that Piel describes as "manifestations of improductive expenditure," I would add the medium of television, which, after all, is the preeminent site for the expenditure of cultural excess in India today. Therefore, I argue that close readings of transgressive excesses on television provide an illuminating framework for the study of the unimaginable communities of nationalism in the dynamic flows of electronic capitalism. As I have shown in my examination of the *Nikki Tonight* controversy, television provides us with the most prominent, and perhaps the few remaining, sites in the national community where an excess of imaginations is constantly produced, consumed, and lost in a flash. In addressing the question "Is there an Indian community of *television?*" in terms of the *Nikki Tonight* controversy, I contend that the expenditure of an excess through a constant process of exposure and erasure—that is to say an exposé—on television is a crucial element in the making and unmaking of the nation as an unimaginable community in postcolonial India.

Conclusion: Is There an Indian Community of Television?

IN 1991, just as the explosion of foreign and domestic satellite television channels began to transform the political, cultural, and economic landscape in India, *Seminar,* an influential academic journal, devoted an entire issue to the problematic of status of the nation as a unified community in relation to the many cultural differences of religion, region, language, class, caste, and so on. This is how D. P. Pattanayak opened the debate on Indian nationalism in his introductory essay, entitled "The Problem":

> In a pluri-cultural society identity assumes a depth and complexity which is difficult to unravel. Linguistic, regional and religious identities are so intertwined with the social and political that in order to understand one, we have to remove the layers of others. Take for example a person from Chamba in Himachal Pradesh. The regional identity is hidden under three layers of linguistic identity, Chambali, Pahadi, and Hindi. The social class and caste identities as well as that of religious faith intersect these identities. *All these identities complement one another. They are not in adversarial relationship.*[1]

A. K. Ramanujan would have perhaps agreed that the way Pattanayak resolves the problematic of Indian nationalism as a community through a comforting myth—that all the identities are complementary and not adversarial—is similar to how an Irishman is supposed to have resolved the troublesome riddle of the trousers.[2] When asked whether trousers were singular or plural, he said, "Singular at the top, plural at the bottom."[3] This quaint resolution may satisfy those like Pattanayak who draw on what is commonly known as the "unity in diversity" model of Indian nationalism. This model of nationalism is celebrated on Indian television in several national integration messages that are regularly seen on Doordarshan and are now also being shown on many of the domestic and foreign satellite and cable channels. For many decades now, government agencies, such as the Films Division in the Information and Broadcasting Ministry, have created short animations, running for about two to three minutes, that have been periodically shown in cinema theaters and also broadcast on Doordarshan.

One such animation invites the Indian viewer to imagine the national community as a great old cultural tree. It represents the nation's collective sense of identity as the root metaphor of the cultural tree and depicts the different regional and local imaginations of community as subsidiary branches of the pan-Indian root-trunk. Whether in the realm of political rhetoric, in the popular imagery of mass media, or in the polemical world of academia, the unity-in-diversity model, by popular consent, is the hegemonic mode for nationalist imaginations in postcolonial India; be it in secular or religious terms, state-sponsored or commercially based productions.

In one of the more popular messages of national integration broadcast on Doordarshan, *Mile Sur Hamra,* the unity-in-diversity model of nationalism is celebrated, once again, quite literally. It features well-known artists in classical music, dance, and cinema, who sing and dance to a song that is played in a variety of Indian languages. The essential theme of the message, as it is performed in different languages, speaks about how one's native tune merges into another's to create a collective tune that belongs to one and all. As the tunes flow from all directions, the song goes, they blend into the oceans, from which they evaporate into the clouds, and then fall gently as raindrops on everyone in the nation.

Another national integration ad, also airing on Doordarshan, features cricket superstars like Sunil Gavaskar and Kapil Dev and the nationally renowned sports figures P. T. Usha and Prakash Padukone carrying a torch across a variety of locations in India, and ending the message with the slogan "Mera Bharat Mahan" (My India Is Great). Yogesh Chandra, a government official who headed the committee that produced this message for the Nehru centenary celebrations in 1989, declared that the "Mera Bharat Mahan" campaign engendered a new sense of patriotism among television viewers. He proudly proclaimed: "If Kapil Dev hits a sixer somebody [in the cricket stadium] raises a banner saying *Mera Bharat Mahan.* Go to a village in Bulandshahr and you see it stenciled on the back of a tractor. Even when that opposition woman heckled the prime minister in the Rajya Sabha [the upper house in Parliament] she said *Mera Bharat Mahan.* It has sunk into the subconsciousness of the people."[4]

In the 1990s, many of the private satellite and cable channels, such as Zee TV, Sony TV, and STAR TV, started challenging Doordarshan's monopoly on messages of national integration by airing what Sivanti Ninan describes as "warm, emotive spots of Indian cricketers in a variety of headgear from the Muslim's fez to the farmer's turban."[5] The message being delivered to the viewers through these spots was that the satellite and cable channels, although privately owned and operated, were as much an integral part of the national community of Indian television as was the state-sponsored network Doordarshan. In 1997, when India celebrated the fiftieth anniversary of the nation's independence, a private Indian company called Bhartbala Productions launched a national integration campaign with a song entitled "Ma Tujhe Salaam" (Mother, I Salute You), which invokes the slogan "Vande Mataram," which was used as a rallying cry in freedom struggle movement.

"Bande Mataram," as the slogan is pronounced in Bengali, is the title of a song that was written in 1875 by the well-known Bengali novelist and nationalist Bankim Chandra Chatterjee to counter the British national anthem "God Save the Queen," which to his consternation was being played in India by the colonial administrators during the 1870s. Bankim wrote the song in a mixture of Bengali and Sanskrit, creating a prayer in which Bharat (India) is saluted as the "mother" of the nation. The song was later included in Bankim's novel *Anandmath*, which was also serialized in his magazine, *Bangdarshan*, between 1880 and 1882.

In 1997, when Bharatbala Productions launched its version to celebrate the fiftieth anniversary of India's independence from the British, it featured the celebrated music director A. R. Rehman, who performed the "Vande Mataram" tune in Raga Desh Malhar. In Indian classical music, a *raga* is defined as a traditional pattern of notes around which a piece of music can be composed and improvised. Raga Desh Malhar belongs to a family of ragas that depict the flow of rainwater in the monsoon season in India. Rehman improvised upon the traditional rendition of this raga and sang it with an upbeat tempo and a quick rhythm to attract young audiences to the nationalist message. Declaring his intent to "rekindle patriotic fervor" among the youth, G. Bharat, the producer of this privately sponsored national integration campaign, released videotapes, cassettes, and CDs of Rehman's contemporary rendition of "Vande Mataram," which became an instant hit across the country. However, since the government of India did not support this commercial use of "Vande Mataram," a parliamentary committee was set up to investigate the matter. When the forty-member parliamentary committee heard Bharatbala Productions' creative use of the song, "they fell into silence and endorsed it."[6]

After Bharat and his team received the parliamentary committee's endorsement, commercial sponsorship came more easily, and the "Vande Mataram" campaign was telecast on a variety of foreign and domestic channels such as BBC, CNN, Doordarshan, STAR TV, Sun TV, and Zee TV. Elated by the response, Bharatbala productions approached Indian cinema's foremost playback singer, Lata Mangeshkar, to chant the "Vande Mataram" song. To go beyond celebrity voices, Bharatbala productions announced that the campaign's long-term goal was to use a chorus of unknown Indians to sing the song in a variety of languages such as Hindi, Tamil, Kannada, Malayalam, and Telugu, to name a few. Unveiling an ambitious plan to market nationalist ideas as commercial products, Bharat declared, "Vande Mataram only means Mother, I bow to thee. We plan to internationalise those magic words extending their emotion to cover Mother Universe. It becomes a global idea for the human planet, a new attitude."[7]

Of course, productions of "Vande Mataram" have been very common in Indian cinema and on national television. For several decades now, radio audiences and television viewers in India have been greeted by "Vande Mataram" every morning, since All India Radio and Doordarshan use it as their signature tune at the beginning of their daily broadcasts. In a recent Bollywood film, *Kabhi Khusi Kabhi Gam* (also known as K3G for short), the song "Vande Mataram" travels with Hrithik Roshan's character to the streets of London in England, where it is played as an upbeat song-and-dance sequence with a group of Indian and British dancers performing in the background. Although such renditions of the nationalist song in diasporic contexts may be relatively new, this phenomenon has also been supported by the rise of domestic and foreign satellite channels, so that private companies such as Bharatbala Productions have used Vade Matram to instill a new sense of nationalist imagination in Indian youth culture.

Surveying "the good, bad, and ugly" contributions of satellite and cable channels in the national community of Indian television, the media analyst Iqbal Malhotra argues that "TV has brought multiple choices in entertainment to the drawing rooms of millions of viewers across the country and established the viewer as the king."[8] At the same time, he worries that "TV channels are mushrooming in an unstructured manner because of the glamour and high hopes of broadcasters."[9] The television producer Nalini Singh agrees that "TV is a lever in a moving society along various lines of thought," but she cautions that "there is an unevenness in the manner in which the entertainment is reaching the viewer in rural and urban areas."[10] Singh argues against a "let's have fun" attitude toward television and calls for "serious and thought-provoking programming" in India.[11]

In this concluding chapter, I survey some of the many views expressed by media scholars, critics, and policy-makers about the role of television in creating, sustaining, and disrupting the status of nationalism in postcolonial India. I situate this contentious debate in relation to the key themes raised in the preceding chapters by revisiting the five question of an Indian community of television with which I began:

Is there an Indian community of television?
Is there *an* Indian community of television?
Is there an *Indian* community of television?
Is there an Indian *community* of television?
Is there an Indian community of *television*?

Is *There an Indian Community of Television*

To the first question, "*Is* there an Indian community of television?" one might say that there *was* an Indian community of television in the second half of the twentieth century, but at the dawn of the twenty-first century, there isn't one anymore. In this changing context, some would argue that the question of an Indian community of television is a thing of the past. This is a view that is encapsulated—with exaggerated effect—in the following blurb that appears on the back cover of Sivanti Ninan's book *Through the Magic Window: Television and Change in India:*

> Television in India started in 1959 but for long remained . . . a slumbering giant muzzled by government control. The satellite revolution of the 90s changed all that. . . . On tap through the day—and night—were slick soaps, uninhibited talk shows, technical dazzlers from Hollywood, news and views from around the world, sports and music, and plenty of blood and gore. This cultural mishmash has left an essentially conservative society bewildered— and sometimes furious—as it gamely tries to cope with the deluge of information and hard sell.[12]

While Ninan describes the "cultural mishmash" of satellite television as a bewildering revolution for Indian viewers, others who take a more evolutionary perspective find that India never changes. This view is evident in the following observation by P. V. Indiresan at the Indian Institute of Technology, Delhi. Exasperated by the slow pace of change in vital sectors of telecommunications, Indiresan writes:

> Since Vedic times, the Indian attitude has been one of suspicion toward all forms of communication. The rishis were sceptical [*sic*] if not hostile, to the written word, and they prohibited the spread of knowledge. In the historic period, the country was averse to keeping systematic records, with the result that we are indebted to foreign sources for knowledge of our own past. In more recent times, we did not take to the use of paper and the printing press till long after they had become commonplace elsewhere. Now there is a veritable war being waged against the introduction of computers. There is something in the Indian psyche which makes us obsessively afraid of sharing or divulging information, with the result that we instinctively draw back from taking any step toward better communications. In fact, we prefer the pains of ignorance to the risks of spreading knowledge.[13]

Is There an Indian Community of Television?

In response to the second question, "Is there *an* Indian community of television?" one might say that there was *an* Indian community of television during the 1960s and 1970s or even the 1980s when Doordarshan was the only national network available to viewers across the nation. But since the arrival of STAR TV in 1991 and the subsequent growth of satellite and cable channels in a variety of languages, it is no longer possible (if it ever was) to speak of *an* Indian community of television. One could argue, as Ramanujan does, that there is a heterogeneity of communities in India that is constituted by "Great and Little Traditions, ancient and modern, rural and urban, classical and folk."[14]

Others might respond to the question of *an* Indian community of television a bit differently by arguing that transcending the apparent diversity of communities there is an overarching unity. Arguing that "the Hindu world image" always reasserts itself as the "community unconscious" in India, the eminent psychoanalyst Sudhir Kakar writes:

> Shared by most Hindus and enduring with remarkable continuity through the ages, the Hindu world image, whether consciously acknowledged and codified in elaborate rituals, or silently pervading the "community uncon-

scious," has decisively influenced Indian languages as well as ways of thinking, perceiving and categorizing experience. This image is so much in a Hindu's bones he may not be aware of it. The self-conscious efforts of westernized Hindus to repudiate it are by and large futile, based as they are on substantial denial.[15]

One the other hand, those who find Kakar's worldview too Hindu-centric to articulate diverse imaginations of gender, class, caste, age, religion, region, and language in the national community would perhaps argue for a more secular Indian way. Proponents of this view point to numerous precedents in precolonial, colonial, and postcolonial times when Indians have welcomed strangers, travelers, rivals, enemies, invaders, and colonizers and have successfully internalized the "otherness" of these external threats into the self-definition of their age-old community.

In *Hype, Hypocrisy and Television in Urban India,* Amrita Shah finds that such sentiments are commonly expressed by prominent Indian journalists who believe that "once the dust had settled down, once the Stars, the CNNs and MTVs of the world had been forced to adapt and restructure to include local language, local events and locally sourced programming . . . the Western beast had been tamed [and] Indians had held their own."[16] Among the examples she quotes is the following passage from a column written by the bureaucrat-turned-politician Mani Shankar Aiyar in a 1994 issue of *Sunday* magazine: "The English tried to capture our minds with the English language; we've now captured the minds of the English with Vikram Seth . . . the West tried to teach us disco; we riposted with disco *dandiya.* The West tried to give us rap. We've served them back with Baba Sehgal. Why on earth should we scare ourselves silly over cultural imperialism?"[17]

Is There an Indian *Community of Television?*

The third question, "Is there an *Indian* community of television? " might elicit answers such as "What we see in India is nothing special to India; it is nothing but preindustrial, pre–printing press, face-to-face, agricultural, feudal."[18] At the very core of such assertions about India's status as a postcolonial nation rests a set of presumptions about the process of modernization and the ideological triumph of Western liberalism during the post–World War II era. There are many who would agree with Walt Rostow, the Harvard economist, presidential adviser, and Cold War strategist who outlined a path to development for postcolonial countries that ultimately concludes with the emergence of a social order that bears a striking resemblance to that of the United States.[19]

Taken as the norm, the "American way" is seen as the standard of the capitalist mode of "development." Many nationalist elites in India, who have for years tried to steer a nonaligned course by maintaining ties with both the First and Second World powers in the West and East, respectively, now find themselves irresistibly drawn to the economic models, investment strategies, and development projects launched by the Western bloc during the Cold War era.[20]

However, there are others who firmly believe that "there is an *Indian* way, and it imprints and patterns all things that enter the continent; it is inescapable, and it is Bigger Than All of Us."[21] On the other hand, some journalists, such as Prasun Sonwalker of the *Times of India,* caution that the pendulum may be swinging the other way now, as imprints of Indian television can be found all across the subcontinent.[22] They argue that the expansion of the Indian television industry in recent years has the makings of "little cultural/ media imperialisms" in South Asia. Quoting Sanjay Baru of the *Business Standard* and Amit Baruah of the *Hindu,* Sonwalker writes:

> In Pakistan, a paranoid military Government is in a rush to create a "media deterrent" against Indian television channels, which are widely seen by the elite of the country. Zee TV and Sony have penetrated into the upper middle class Pakistani homes as never before. . . . As Pakistan's information minister, Javed Jabbar, puts it, "I am concerned about the influence of Indian satellite television on our people."[23]

Is There an *Indian* Community *of Television?*

The fourth question, "Is there an Indian *community* of television?" may revive longstanding academic debates about whether Indians think of themselves as a "community" at all. When people compare Indian imaginations of "community" with Western notions of the same, there are hotly contested arguments for and against it. "Some lament, some celebrate India's unthinking ways."[24]

Kalpana Ram[25] outlines the cultural potency of these two views—lament and celebration—in Walter Goldschmidt's[26] and Stephen Tyler's[27] anthropological introductions to India. In a foreword to A. R. Beal's book on village life, Goldschmidt laments India's nonmodern sense of community thus:

> Of the world's major civilizations, the Indian is perhaps the most difficult for Westerners to comprehend. The formal structure of castes with their apparent fixity, the strong sense of hierarchy supported ritually by the concept of pollution, the religious belief in reincarnation, as both a justification of one's present condition and as a motivation for proper conduct, are all alien to our way of thinking. . . . Early Western descriptions, generalized and remote from daily realities, have exaggerated rather than promoted understanding. Indian scholars steeped in their own traditions and viewpoints, offering descriptions of the system as a core theoretical and theological construct, have hardly been more helpful.[28]

On the other hand, Tyler finds postmodernist celebrations of dialogism and polyvocality in Indian ways of thinking about community. He argues:

> On the whole, Indians are not seriously affected by the kind of cognitive dissonance (the mental discomfort of holding contradictory views) that either paralyses Westerners or galvanizes them into action directed at bringing consistency back into their mental lives. Among Indian intellectuals, this disarming of mental dissension promotes a jumbled and uncritical eclecticism. . . . Facts and ideas are indiscriminately garnered and jealously sequestered in whatever nook and cranny and regularly trotted out in relevant support of any conceivable doctrine.[29]

Is There an Indian Community of Television?

In posing the fifth and the final question, "Is there an Indian community of *television?*" I have tried to focus attention on the role that television plays in articulating diverse imaginations of the nation as a community. However, to the many critics of television, this final question may in fact be a nonquestion. They argue that television does not (and should not) play any role in answering onerous questions about "nationalism," "imagination," and "community" in India. Television, they argue, is merely an entertainment medium and is not conducive to answering serious epistemological, ontological, and theological questions raised in debates about the status and the nature of the national community. At best, they concede, television serves as an illusionary device for its audiences by taking their attention away from growing concerns about national development, communal violence, familial harmony, and individual rights and responsibilities in the real world of everyday life.

In *The Impact of Television Advertising on Children,* Namita Unnikrishnan and Shailaja Bajpai confess to "a growing sense of unease at what has been happening in the world of television."[30] They write:

> TV advertising is imposing an image of life that is completely alien to the vast majority of Indian children. Many children are beginning to believe that the India and the Indians they see in TV ads are the only ones worth emulating and learning from. . . . As a result, material aspirations are reaching unrealistic heights. Development has never appeared so disjointed. While consumerism is spreading like wildfire, access to the basics of life remains a serious problem for many.[31]

However, there are many in the world of television who share the unease expressed by Unnikrishan and Bajpai but remain convinced about television's potential as a medium of education, information, and entertainment. Even as they are dismayed by the growing commercialization of the medium, they are hopeful that the state-sponsored network, Doordarshan, will ultimately fulfill its public service mission as the chief unifier of the nation. In an ethnographic study of television viewers in Danawali village in the state of Maharashtra, Kirk Johnson argues that the commercial agenda of Indian television in the 1990s "has little if any resemblance to" the cherished goals of national development in the service of which the medium was brought to the country in 1959.[32] Scathingly critical of the urban middle class for enabling Indian television to be "wooed away from its humanitarian goals," Johnson writes: "It is this class which owns and operates most of the television industry in India. And it is also this class which is transmitting its own values, principles and opinions to the rest of India. And though Doordarshan has attempted to stay true to its original goals it has had to succumb to the market forces to survive."[33]

However, even among the advocates of Doordarshan's cherished goals of national integration and rural development, there are those who believe that television is too important a medium to be left in the hands of the state-sponsored network. They point to the thirty years of bureaucratic mismanagement in Doordarshan, and they consider the competition from foreign and domestic channels to be the only viable alternative for articulating diverse imaginations of the rural and the urban, the rich and the poor, the hegemonic and the marginal in postcolonial India. Shailendra Shankar, the officer in charge of Doordarshan at the time of its launch in India—who later retired

as its director general—envisions a tremendous opportunity for Indian viewers in the growing competition between the state-sponsored network and
the commercial channels on satellite and cable television: "Hundreds of
channels will offer [viewers] a tremendous variety of programmes to choose
from. They will virtually have the world at the door-step. But with their
gaining experience in television-viewing, they will pick only the very best.
And that indeed would be good for television in India as a whole."[34]

Is There an Indian Community of Television?

By critically engaging with the contentious debate over diverse imaginations
of identity and difference in relation to the foregoing five questions, I have
attempted to provide an alternative framework with which to interrogate
the articulation of nationalism to electronic capitalism in Indian television.
This alternative view not only acknowledges the obvious—that India is a
nation that is constituted as a unity through its diversity—but also posits
that the interplay between unity and diversity is, however difficult it may be
for us to imagine, at once conflicting and complementary.

The conflict and complementarity of the various perspectives on the
unity and diversity of Indian nationalism can be best understood through
the principle of self-reflexivity, which, as Ramanujan argues, refers to the
creative process of constantly generating new forms out of old ones.[35] In the
cultural process of regeneration, self-reflexivity takes many forms: "awareness
of the self and other, mirroring, distorted mirroring, parody, family resem-
blances and rebels, dialectic, antistructure, utopias, dystopias, the many iro-
nies connected with these responses."[36]

When we examine the various incarnations of *Vande Mataram* through
the lenses of self-reflexivity and intertextuality in India, what is merely suggested in one text often becomes a central element in a repetition or imitation
of it. As Ramanujan puts it, "mimesis is never only mimesis, for it evokes the
earlier image in order to play with it and make it mean other things."[37] Therefore, when the "same" texts appear in different contexts, "we cannot dismiss
them as interlopers and anachronisms, for they become signifiers in a new
system: mirrors again that become windows."[38]

The reordering of texts in a cultural simultaneity of conflict and complementarity enables us to understand how "every new text within a series con-
firms yet alters the whole order ever so slightly, and not always so slightly."[39]

Ramanujan tells us that the question of self-reflexivity and intertextuality in Indian ways of thinking was posed to him, in very personal terms, in the image of his father. He recalls:

> My father's clothes represented his inner life very well. He was a south Indian Brahmin gentleman. He wore neat white turbans, a Sri Visnava caste mark (in his earlier pictures, a diamond earring), yet wore Tootal ties, Kromentz buttons and collar studs, and donned English serge jackets over his muslin *dhotis* which he draped in traditional Brahmin style. He often wore tartan-patterned socks and silent well-polished leather shoes when he went to the university, but he carefully took them off before he entered the inner quarters of the house.
>
> He was a mathematician, an astronomer. But he was also a Sanskrit scholar, an expert astrologer. He had two kinds of exotic visitors: American and English mathematicians who called on him when they were on a visit to India, and local astrologers, orthodox pundits, who wore splendid gold-embroidered shawls dowered by the Maharajah. I had just been converted by Russell to the "scientific" attitude. I (and my generation) was troubled by his holding together in one brain both astronomy and astrology; I looked for a consistency in him, a consistency he didn't seem to care about or even think about. When I asked him what the discovery of Pluto and Neptune did to his archaic nine-planet astrology, he said, "You make the necessary corrections, that's all." Or, in answer to how he could read the *Gita* religiously, having bathed and painted on his forehead the red and white feet of Visnu, and later talk appreciatively about Bertrand Russell and even Ingersoll, he said, "The Gita is part of one's hygiene. Besides, don't you know, the brain has two lobes?"[40]

Ramanujan finds that many "rational" observers have been dismayed and angered by the "inconsistency" and "hypocrisy" they perceive in many Indians, such as his father, who appear to seamlessly switch between the logical rigors of scientific education and the cultural imaginations of archaic traditions. Ramanujan argues that the "inconsistency" or "hypocrisy" in their ways of thinking may be due not to "inadequate education" or "lack of logical rigor" but the fact that they may be using a different logic altogether.[41]

This altogether different logic, which Ramanujan describes, in very personal terms, as exemplified by his father's "Indian" ways of thinking, is precisely what I have sought to address in more general terms by focusing on the uses and abuses of the name of Mahatma Gandhi, who performs the symbolic function of the Father of the Nation in postcolonial India. By de-

constructing the ambivalent status of Gandhian nationalism in my consideration of the unimaginable communities of Indian television, I have sought
to emphasize three important distinctions within what Arjun Appadurai has
called the work of imagination in electronic capitalism as follows.[42]

First, as electronic media networks deliver a variety of images, ideas,
ideologies, and commodities from around the world into television households, imagination has ceased to be the exclusive preserve of nationalist elites,
cult figures, and messianic leaders who seek to implant their political visions
of nationalism and of revolutionary change in the everyday lives of ordinary
people. As the uneven flows of television cultures becomes integral to the
practice of everyday life, viewers are induced to embrace hybrid imaginations
of national identity and cultural differences when they make critical decisions
about spiritual and material relations in both the private domain of the home
and in the public domain of the world.

Second, the rapid integration of television into the social fabric of everyday life increasingly brings into question a distinction that Appadurai describes as the disjuncture between imagination and fantasy in electronic
capitalism.[43] There is a long-established tradition of media criticism—cutting
across ideological divides such as "left-wing" and "right-wing" in academic
and nonacademic circles—that views television as escapist fantasy and fears
that the advance of electronic capitalism through the commercially satu-
rated media networks negatively impacts the development of independent
thought, expression, and imagination among young and old viewers alike.
However, there is a growing body of scholarship—most notably in the field
of cultural studies—that does not embrace such deterministic views about
the power of mass media and focuses instead on the creative ways in which
viewers use (and often abuse) the power/knowledge of television to assert
their own sense of identity, difference, nationality, and transnationality. Stuart Hall succinctly summarizes the latter approach—the cultural studies, or
culturalist, approach—to understanding the work of imagination in the
dynamic relationship between television and the everyday life of viewers as
follows.

> In its different ways, it [cultural studies] conceptualizes culture as inter-woven
> with all social practices; and those practices, in turn, as a common form of
> human activity; sensuous human praxis, the activity through which men and
> women make history. It is opposed to the base and superstructure way of
> formulating the relationship between ideal and material forces, especially
> where the base is defined by the determination by the "economic" in any

simple sense. . . . It defines "culture" as both the means and values which arise amongst distinctive social groups and classes, on the basis of their given historical conditions and relationships, through which they "handle" and respond to the conditions of existence; *and* as the lived traditions and practices through which those "understandings" are expressed and in which they are embodied.[44]

Finally, to emphasize a third distinction of electronic capitalism, which Appadurai describes as the difference between the individual and collective senses of imagination, I favor the culturalist approach of understanding media as "the staging ground for action" over the deterministic view of media as escapist fantasy.[45] To speak of imagination as an individual phenomenon is to focus on the flowering of the mental faculties of thought and the independent expression of ideas by a gifted individual (such as an author, an artist, or a poet). In the postelectronic world, as television networks provide a large number of channels for the mass communication of ideas and the mass experience of pleasures, imagination has increasingly become a collective property.

To foreground the potential for collective action (political, aesthetic, ethical, or even unethical) through the work of imagination is not to suggest that television viewers in India, or the Indian television networks, are autonomous agents unregulated by the ideological powers of electronic capitalism. Rather, it is an attempt to rearticulate the ideology of electronic capitalism that has now reached every corner of the globe, in terms of a multiplicity of sites where television has considerable influence over diverse imaginations of nationalism, transnationalism, and translocalism. What is at play in this redefinition is a regulated free agency; the collective agency of viewers who are free to be at home with their television sets, even as they are regulated by the ideological constraints of electronic capitalism.

The redefinition of agency as a regulated freedom to collectively engage with the medium of television has a significant bearing on the process of imagining nations as communities in electronic capitalism. Even as we move away from the traditional equation of community with the geographical boundaries and the physical places that constitute the nation space, we must recognize that nationalism is not something that will fade away in the near future. Commenting on the cultural tensions that secular and religious authorities around the world have been forced to encounter due to the rapid deterritorialization of their communities through television, Chris Barker writes: "Diverse cultures which had once been considered 'alien' and remote are becoming accessible today (as signs and commodities) via our televisions, radios and shopping centers. As a consequence we may choose to eat 'Indian,' dress 'Italian,' watch 'American' and listen 'African.'"[46]

As Barker rightly argues, in these diverse contexts of electronic capitalism, a central issue that remains contentious is the question of nationalist imagination on television. By focusing on the ambivalent articulation of identity and difference through the paradoxical formulation of unimaginable communities in Indian television, I have also tried to address the question of why conventional theories of nationalism are not adequate to make sense of changes engendered by the dynamic flows of electronic capitalism. In doing so, I have neither attempted to reveal a universal logic of identity nor sought to revel in the infinite play of differences. Instead, I have relied upon a paradoxical formulation of unimaginable communities that at once defines and defies the limits of national identity and cultural differences in Indian television.

In positing the notion of unimaginable communities to address the interplay of identity difference in Indian television, I may be guilty of constructing conceptual hybrids, theoretical mongrels, and methodological medleys, some of which may stand the scrutiny of academic rigor while others may not. If I have chosen to move away from some of the more conventional theories and methodologies of cultural criticism and invoke more archaic notions of the same, it is with the awareness that the privileged position one assumes as an elite critic is, as Gayatri Spivak argues, as much a loss as a privilege.[47]

As Ashis Nandy argues in the context of cultural criticism in India, the aim of research should not be to adjust the Indian experience to fit the existing cultural theories.[48] Rather, what the cultural critic must attempt is "to make sense of some of the categories of contemporary knowledge in Indian terms, and put forth a competing theory of culture in postcolonial discourse."[49] In this study of the unimaginable communities of Indian television, if I have been able to partly follow the strategy for cultural criticism that Nandy advocates, I would consider this effort as being worth the time; and the time as being well spent.

```
┏┓
┗┛
```

NOTES

Introduction

1. See Chatterjee, *Nationalist Thought and the Colonial World*, and *Nation and Its Fragments*, and Anderson, *Imagined Communities*.

2. See, for instance, Mankekar, *Screening Culture, Viewing Politics*, and Rajagopal, *Politics after Television*. Mankekar and Rajagopal, whose books were published in 1999 and 2000, respectively, conducted most of their fieldwork on Doordarshan's national programming before the advent of satellite television in India. Therefore, I find that their books, while providing crucial insights into the growing influence of television in India, are unable to adequately attend to the changing status of the national network in relation to the sweeping transformations of the transnational and translocal media networks since the early 1990s. See my review of Mankekar's *Screening Cultures, Viewing Politics*, and Rajagopal's *Politics after Television*, in *Screen*.

3. Kellner, *Persian Gulf TV War*, 110.

4. Ibid., 111.

5. Quoted in Jain, "Hooked on War," 38.

6. Ibid.

7. For an overview of the transformations of the satellite television industry in the early 1990s, see Bhatt, *Satellite Invasion of India*, Gupta, *Switching Channels*, Ninan, *Through the Magic Window*, and Saksena, *Television in India*.

8. "How Cable TV Began and Spread in India."

9. Ibid.

10. "Doordarshan Today."

11. Pendakur and Kapur, "Think Globally, Program Locally."

12. Kumar and Curtin, "'Made in India.'"

13. The Sachs report is cited in Watson, "Asia's Rising Star."

14. Ibid.

15. Zee TV, *Annual Report*, 1996.

16. Pendakur and Kapur, "Think Globally, Program Locally," 202.

17. Ibid.

18. Ibid.

19. Sonwalker, "India: Makings of Little Cultural/Media Imperialism?"

20. Ibid., 511.

21. In 1997, Rajat Sharma and forty-three others affiliated with *Aap Ki Adalat* had resigned in protest over what Sharma claimed was excessive interference in the programming content of the news show by Vijay Zindal, the chief executive officer at Zee. After quitting Zee, Sharma formed his own production company in New Delhi.

22. Jeffery, *India's Newspaper Revolution.*

23. Lahiri, "Eenadu Chief Claims New Channel Has Penetrated All 1.3 M Cable Homes in A.P."

24. Anand, "Sun TV."

25. Lahiri, "Eenadu Chief Claims New Channel Has Penetrated All 1.3 M Cable Homes in A.P."

26. Ninan, "Channel after Channel."

27. Anderson, *Imagined Communities*, 7.

28. Ibid.

29. Ibid.

30. Ninan, "Indelible Images."

31. Ibid. Lal Bahadur Shastri was India's second prime minister and was sworn into office after the passing away of Pandit Jawaharlal Nehru in 1964. Shastri served as prime minister for two years until his death in 1966.

32. Derrida, *Specters of Marx*, 6.

33. Ibid.

34. In this book, I follow the conventional practice of using the name "Gandhi" along with the honorary title "Mahatma" (the Great Soul) instead of referring to the proper name M. K. Gandhi.

35. Khosla, "Channel Surfing."

36. Menon, "Mahatma Brings Alive the Early Days and Little Known Facts of Gandhi as a Child."

37. Farrell, "Now Sir Humphrey Gets to Say 'Yes Minister' in Hindi."

38. Quoted in Nair, "Thumpingly Successful Viji Thampi."

39. Ibid.

40. Nasta, "Trouble at Bombay Doordarshan."

41. Bareth, "Gandhi Fashion Image Row."

42. Bhatia, "Cries of Cultural Imperialism Examined."

43. Ibid.

44. Ramanujan, "Is There an Indian Way of Thinking?"

45. Ibid., 40.

46. Ibid.

Chapter 1: From Doordarshan to Prasar Bharati

1. Quoted in "Fast Forward to a Slice of History," *Hindu*, 2001.

2. UNESCO, *Television*, 105.

3. Chatterji, *Broadcasting in India.*

4. Kumar and Chandiram, *Educational Television in India.*

5. Ibid.

6. Ibid.

7. Ministry of Information and Broadcasting, *Radio and Television* (also known as the Chanda Committee Report), paras. 778, 779, 780.

8. Rao, personal interview.

9. Ibid.

10. Ninan, *Through the Magic Window.*

11. Luthra, *Indian Broadcasting.*

12. For a detailed account of the reasons for, and the consequences of, Indira Gandhi's proclamation of "Emergency," see Bahl, *Indira Gandhi,* and Dhar, *Indira Gandhi.*

13. Luthra, *Indian Broadcasting,* 274–75.

14. Basu, *Introduction to the Constitution of India.*

15. Ibid., 392.

16. Luthra, *Indian Broadcasting,* 434.

17. Mitra, *Television in India.*

18. Ministry of Information and Broadcasting, *Major Recommendations of the Working Group on Autonomy for Akashvani and Doordarshan* (also known as the Verghese Committee Report).

19. Quoted in Shah, *Hype, Hypocrisy and Television in Urban India,* 11.

20. Ministry of Information and Broadcasting, *An Indian Personality for Television* (also known as the Joshi Committee Report), 7.

21. Ibid., 18.

22. Quoted in Unger, "TV Comes to India," 25.

23. Singhal and Rogers, "*Hum Log* Story in India," 75.

24. Ibid., 101.

25. Rajagopal, "Rise of National Programming," 92.

26. Mitra, *Television in India,* 18.

27. Ibid., 19.

28. Rajagopal, "Rise of National Programming,"92.

29. Ibid.

30. Rajagopal, *Politics after Television.*

31. Ibid., 92.

32. Ibid., 93.

33. Ibid., 28.

34. For critical analyses of the Ramjanmabhoomi/Babri Masjid controversy, see Gopal, *Anatomy of a Confrontation,* and Nandy et al., *Creating a Nationality.*

35. Mankekar, *Screening Culture, Viewing Politics.*

36. "Prasar Bharati Bill," 104.

37. For a more detailed discussion of the Narasimha Rao government's liberalization policies in the early 1990s, see chapter 4.

38. Verma, "Air Waves."

39. Cited in Ninan, "History of Indian Broadcasting Reform," 13.

40. Cited in Swami, "Autonomy in Prospect."

41. Ibid.

42. Cited in Ninan, "History of Indian Broadcasting Reform," 14.

43. The organizational structure of Prasar Bharati and the current composition of the membership of its board can be found online at the website of Prasar Bharati: http://www.ddindia.net/dd prasarbharati.html.

44. Reuters Textline, "Indian Newcomers Find the Going Tough."

45. For an overview of the DTH and the Pay TV phenomena, see Wanwari, "Pay TV Conundrum" and "Direct to Home Broadcasting Sees Light of Day."

46. Jha, "Prasar Bharati Fate Hangs in Balance."

47. Quoted in "'I do not have faith in Prasar Bharati': Minister Declares."

48. For an overview of the CAS debate, see Mittal, "Cable TV 'CAS'T Away."

49. The structure and functions of the Broadcasting Authority of India, as proposed in the Broadcasting Bill of 1997, can be found online at the website of the Network of Women in Media, India: www.nwmindia.org/Law/Bare acts/broadcastbill.htm.

50. Bhabha, *Location of Culture.*

51. Bhabha, "DissemiNation."

52. See Derrida, *Dissemination,* and "Differance."

53. Bhabha, "Preface: Arrivals and Departures."

54. Ibid., viii.

55. Ibid., ix.

56. Derrida, *Specters of Marx.*

57. Ibid.

58. Bhabha, "Preface: Arrivals and Departures," xi.

59. Ministry of Information and Broadcasting, *Radio and Television,* 210.

60. Ministry of Information and Broadcasting, *Major Recommendations of the Working Group on Autonomy for Akashvani and Doordarshan*

61. Ministry of Information and Broadcasting, *An Indian Personality for Television,* 18.

62. Ministry of Information and Broadcasting, *The Nitish Sengupta Committee Report on Prasar Bharati.*

Chapter 2: At Home, In the World

1. Chakrabarty, *Provincializing Europe,* and Anderson, *Imagined Communities.*

2. Chakrabarty, *Provincializing Europe,* 175.

3. Ibid., 173.

4. For the early history of television manufacturing in India, see Kumar, *Television Industry in India.*

5. *Illustrated Weekly of India,* November 25, 1973, 3.

6. Ibid.

7. Spigel, *Make Room for TV.*

8. Ibid., 122.

9. Kumar, *Television Industry in India.*

10. *Illustrated Weekly of India,* July 9, 1972, 72.

11. Ibid.

12. Ibid.

13. Ibid.

14. *Illustrated Weekly of India*, November 5, 1972, 72.

15. Ibid.

16. Williams, *Television*.

17. *Illustrated Weekly of India*, November 5, 1972, 72.

18. *Illustrated Weekly of India*, June 24, 1973, 4.

19. *Illustrated Weekly of India*, July 4, 1973, 58.

20. *Illustrated Weekly of India*, August 26, 1973, 65.

21. Spigel, *Make Room for TV*, 167.

22. Ferraro, "Popular Reaction to Bombay TV," 30.

23. Kumar, *Television Industry in India*.

24. *India Today*, July 16–31, 1977. Inside front cover.

25. *India Today*, September 16–30, 1977, 11.

26. *India Today*, November 1–15, 1977, 18.

27. Ibid.

28. *India Today*, December 16–31, 1977, 78.

29. Ibid.

30. Ibid.

31. Kumar, *Television Industry in India*.

32. Razzaque, "Color Television Development at CEERI."

33. *India Today*, December 16–31, 1979, 41.

34. Ibid.

35. Guhathakurta, "Electronics Policy and the Television Manufacturing Industry," 847.

36. Kumar, *Television Industry in India*, 76.

37. Ibid., 45.

38. Ibid., 109.

39. Razzaque, "Color Television Development at CEERI."

40. *Electronics for You*, November 1982, 45.

41. Ibid.

42. *Electronics for You*, November 1982, 51.

43. *India Today*, August 15, 1985, 65.

44. *India Today*, September 15, 1985, 124.

45. *India Today*, December 15, 1985, 131.

46. *India Today*, December 31, 1985, 191.

47. Ibid.

48. Ibid.

49. Guhathakurta, "Electronics Policy and the Television Manufacturing Industry," 849.

50. Ibid., 850.

51. Ibid., 850.

52. Ibid., 851.

53. *India Today*, January 31, 1990, 66.

54. Guhathakurta, "Electronics Policy and the Television Manufacturing Industry."

55. Ibid., 865.

56. Ibid.

57. *India Today,* August 31, 1990, 78.

58. Ibid.

59. Ibid.

60. Guhathakurta, "Electronics Policy and the Television Manufacturing Industry," 865.

61. Prasad, "Giving the Devil Its Due."

62. *India Today,* February 29, 1996, 85.

63. Ibid.

64. Ibid.

65. *India Today,* 1996.

66. *India Today,* 1996.

67. Quoted in Povaiah, "A Fresh Focus," 38.

68. Srikanth, "BPL Plans to Set Up or Acquire Manufacturing Unit in UK."

69. *India Today,* January 31, 1997, 125.

70. Ibid.

71. Ibid.

72. "Color Television Industry."

73. *India Today,* April 9, 2001, 4–5.

74. Ibid.

75. "Color Television Industry."

76. *India Today,* May 21, 2001, 21.

77. *India Today,* June 4, 2001, 61.

78. Ibid.

79. Ibid.

80. *India Today,* April 2, 2001, 22–23.

81. Ibid.

82. *India Today,* April 16, 2001, 22–23.

83. Ibid.

84. Ibid.

85. Appadurai, "Here and Now."

86. Peters, "Seeing Bifocally."

87. Ibid., 79.

88. Appadurai, "Here and Now," 3.

89. Spigel, *Make Room for TV,* 102.

90. Ibid., 99

91. Ibid., 182.

Chapter 3: Between Tradition and Modernity

1. Halle, "On Teaching International Relations."

2. The full text of "Point Four" can be found in Rist, *History of Development,* 249–50. It reads as follows.

Fourth, we must embark on a bold new program for making the benefits of our scientific advances and industrial progress available for the improvement and growth of underdeveloped areas.

More than half the people of the world are living in conditions approaching misery. Their food is inadequate. They are victims of disease. Their economic life is primitive and stagnant. Their poverty is a handicap and a threat both to them and to more prosperous areas.

For the first time in history, humanity possesses the knowledge and skills to relieve the suffering of these people.

The United States is pre-eminent among nations in the development of industrial and scientific techniques. The material resources which we can afford to use for assistance of other peoples is limited. But our imponderable resources in technical knowledge are constantly growing and inexhaustible.

I believe that we should make available to peace-loving peoples the benefits of our store of technical knowledge in order to help them realize their aspirations for a better life.

And in cooperation with other nations, we should foster capital investments in areas needing development.

Our main aim should be to help the free peoples of the world, through their own efforts, to produce more food, more clothing, more materials for housing, and more mechanical power to lighten their burdens.

We invite other countries to pool their technological resources in this undertaking. Their contributions will be warmly welcomed. This should be a cooperative enterprise in which all nations work together through the United Nations and its specialized agencies whenever practicable. It must be a worldwide effort for the achievement of peace, plenty and freedom.

With the cooperation of business, private capital, agriculture, and labor in this country, this program can greatly increase the industrial activity in other nations and can raise substantially their standards of living.

Such new economic developments must be devised and controlled to the benefit of peoples of the areas in which they are established. Guarantees to the investor must be balanced by guarantees in the interest of the people whose resources and whose labor go into these developments.

The old imperialism—exploitation for foreign profit—has no place in our plans. What we envisage is a program of development based on the concepts of democratic fair-dealing.

All countries, including our own, will greatly benefit from a constant program for the better use of the world's human and natural resources. Experience shows that our commerce with other countries expands as they progress industrially and economically.

Greater production is key to prosperity and peace. And the key to greater production is a wider and more vigorous application of modern scientific and technical knowledge.

Only by helping the least fortunate of its members to help themselves can the human family achieve the decent, satisfying life that is the right of all people.

Democracy alone can supply the vitalizing force to stir the peoples of the world into triumphant action, not only against their human oppressors, but also against their ancient enemies—hunger, misery and despair.

On the basis of these four major courses of action we hope to help create the conditions that will lead eventually to personal freedom and happiness for all mankind.

3. Rist, *History of Development*.

4. Ibid.

5. Ibid.

6. Esteva, "Development," 7–8.

7. Rist, *History of Development,* 73.

8. Ibid.

9. Mundt, "We Can Give the World a Vision of America!"

10. Ibid. Schwoch, "'We Can Give the World a Vision of America,'" provides an in-depth historical analysis of the various plans proposed in the United States for the creation of a global television system during the Cold War era. I thank Schwoch for this unpublished essay.

11. Levey, "Democracy and Freedom Can Be Promoted via Global TV."

12. Hickenlooper, "U.S. Sen. Hickenlooper Cites Facts in Support of Global Television Commission."

13. Nandy, *Traditions, Tyranny and Utopias.*

14. It is unclear who first used the phrase "Third World" to describe the postcolonial nations in the international system of states. While some credit is due the former French president Charles de Gaulle for coining the phrase, other accounts suggest that Lee Kuan Yew, the former prime minister of Singapore, used it first in 1969. However, some scholars suggest that it was coined by the French demographer Alfred Sauvy in *Le Nouvel Obersaveteur* 118, August 14, 1952. Although Sauvy's French term "les tiers monde" translates into English as the "the third world (or the three worlds), he used it to refer primarily to India and China, which, he argued, were leading the struggle against European colonialism. Curt C. Campbell and Thomas G. Weiss argue that Sauvy's use of "tiers monde" was an inspired reference to the concept of "the third estate," which they claim was originally used by Emmmanuel Sieyès in January 1789 to describe "the struggling third force in the French Revolution that was combating the clergy and aristocracy." See Campbell and Wiess, "The Third World in the Wake of Eastern Europe."

15. Frenkel and Frenkel, *World Peace via Satellite Communications,* 55.

16. Ibid., 56–57.

17. Some of the early works that debated the role of Intelsat in facilitating international communications include Galloway, *Politics and Technology of Satellite Communications,* Musolf, *Communication Satellites in Political Orbit,* Pelton, "Key Problems in Satellite Communications," Smith, *Communication Via Satellite,* and Snow, *International Commercial Satellite Communications.*

18. For a detailed discussion of the debates over the structure and charter of Intelsat, see *Communicating by Satellite.*

19. Lerner, *Passing of Traditional Society.*

20. Ibid.

21. Samarajiva, "Murky Beginnings of the Communication and Development Field."

22. Rogers, *Diffusion of Innovations.*

23. One of the earliest "diffusion" studies about the use of media for development in India is Lakshmana Rao's classic analysis of two villages in Andhra Pradesh, "Communication and Development," which uses the terms "Pathooru" (old village) and "Kothooru" (new village). The study contrasts the communication patterns of resi-

dents in "Kothooru," who have access to new media, a new road, and modern ideas, with those of the residents in "Pathooru," who live in a more traditional way.

24. Lenin, *Imperialism.*

25. Etinger, *NAM,* 15.

26. Quoted in ibid., 8.

27. Quoted in Galloway, *Politics and Technology of Satellite Communications,* 123.

28. Donetsky, "Despite Major Difficulties a World Television System Is Possible," 51.

29. Galloway, *Politics and Technology of Satellite Communications,* 130.

30. Wallerstein, *Capitalist World Economy.*

31. Ibid., 5.

32. Nandy, *Intimate Enemy,* and *Traditions, Tyranny and Utopias.*

33. Nehru, *Toward Freedom,* 228.

34. Prakash, "Who Is Afraid of Postcoloniality?"

35. Prasad, "The Evolution of Non-Alignment."

36. Prakash, "Who Is Afraid of Postcoloniality?" 196.

37. Ibid., 197.

38. Chatterjee, *Nationalist Thought and the Colonial World.* See also Chatterjee, *Nation and Its Fragments.*

39. Prakash, "Who Is Afraid of Postcoloniality?" 195.

40. See Nandy, *Intimate Enemy,* and *Traditions, Tyranny and Utopias.*

41. Nandy and Visvanath, "Modern Medicine and Its Non-Modern Critics."

42. Ibid., 171.

43. Nandy, *Intimate Enemy,* 57.

44. Ibid., 183.

45. Illich, *Medical Nemesis.*

46. Sarabhai, *Science Policy and National Development,* 40.

47. Sarabhai wrote on many issues of science and technology, including the use of television for national development, the peaceful exploration of outer space, and Gandhian ideals of nonviolence in science and education. For a sample of Sarabhai's writings on these topics, see *Science Policy and National Development.*

48. Rajadhyaksha, "Beaming Messages to the Nation," 35.

49. For a discussion of why India was considered the most suitable site for this experiment by policy-makers in the U.S. government, see Joshi, *Implications of India/U.S. Satellite Instructional Television Experiment.*

50. Shoesmith, "Footprints to the Future, Shadows of the Past."

51. Ibid.

52. Cited in Mitra, *Television in India,* 32–33.

53. Weidner, *Technical Assistance in Public Administration Overseas,* 203.

54. Schramm and Nelson, *Communications Satellite for Education and Development,* 193–94.

55. Beltran, "Alien Premises, Objects and Methods."

56. Srinivasan, "No Free Launch."

57. Quoted in Pagedar, "Non-Aligned News Pool."

58. International Commission for the Study of Communication Problems, *Many Voices, One World.*

59. Rogers, *Communication and Development.*

60. Gerbner, Mowlana, and Nordenstreng, *Global Media Debate.*

61. Prakash, "Who Is Afraid of Postcoloniality?" 198.

62. Ibid.

63. Intersputnik Corporate Brochure, 2001, available online at: www.intersputnik. com/docs/corporate%20brochures%2001e.pdf.

64. Das, "VSNL Set to Join Intersputnik Bandwagon."

65. Bhabha, "Preface: Arrivals and Departures," x.

66. Prakash, "Who Is Afraid of Postcoloniality?" 198. For a post–Cold War perspective on the "end of history" thesis, see Fukuyama, *End of History and the Last Man* and "End of History?"

67. Prakash, "Who Is Afraid of Postcoloniality?" 198.

68. Ibid., 199.

69. Ibid., 200.

70. Ibid., 201.

Chapter 4: "Gandhi Meet Pepsi"

1. Butalia, "Gandhi Meet Pepsi."

2. Ibid.

3. See my discussion of imagiNation in chapter 1 for a more detailed consideration of the writing and erasure of nationalism in electronic capitalism. Also see Bhabha, "Preface: Arrivals and Departures," and Derrida, *Specters of Marx.*

4. Shenon, "A Race to Satisfy TV Appetites in Asia."

5. Ibid.

6. Kristof, "Satellites Bring Revolution to China."

7. Gargan, "New Delhi Journal."

8. Rettie, "India: Slow Stirrings of a Million Mutinies."

9. Ibid.

10. Ibid.

11. Ibid. A research survey conducted by *Advertising and Marketing* magazine describes the many difficulties that marketing agencies have faced in the 1990s in their attempts to accurately estimate the size of the growing Indian middle class. See Singh, "Muddle in the Middle."

12. Rettie, "India: Slow Stirrings of a Million Mutinies."

13. For a detailed analysis of the momentous changes in the Indian economy that took place in a twenty-month period between August 1990 and March 1992, see Jalan, *Indian Economy.*

14. In November 1993, *Business World* did a cover story on the reentry of Coca-Cola into the Indian markets after a hiatus of sixteen years. See Ghosh, "Marketing a Megabrand."

15. "Cola Companies Battle On."

16. In a survey of the food-processing industry in the 1990s, *Advertising and Marketing* magazine asked the question: "Will Processed and Convenience Foods Alter the Way Indians Eat in a Significant Manner?" For detailed findings of the survey, see Sahai, "Whiff of Promise."

17. Quoted in Sawant, "Making of the Mahatma."

18. Fernandez, "India: Reforms Trigger Clash with Indian Values."

19. Ibid.

20. Ibid.

21. Ibid.

22. Duncan, "India: Hello World."

23. Ibid.

24. Ibid.

25. Ibid.

26. Ibid.

27. Quoted in Fernandez, "India: Reforms Trigger Clash with Indian Values."

28. Quoted in ibid.

29. Quoted in ibid.

30. For an extensive overview of the impact of economic liberalization policies in various industrial sectors during the 1990s, see Joshi and Little, *India's Economic Reforms,* and Srinivasan, *Eight Lectures on India's Economic Reforms.*

31. Anjan Mitra, *India Today,* 1996, 14.

32. There are, of course, many reasons for the marginalization of the Nehruvian and the Gandhian ideals in recent years. In *India: Economic Development and Social Opportunity,* Jean Drèze and Amartya Sen outline three "ideological convictions that have contributed to this neglect" as follows.

(1) The conservative upper-caste notion that knowledge is not important or appropriate for the lower castes; (2) a distorted understanding of Gandhi's view that "literacy in itself is no education"; and (3) the belief, held in some radical quarters, that the present educational system is a tool of subjugation of the lower classes, or the vestige of the colonial period. (p. 111)

33. Quoted in Uniyal, "India—Economy."

34. Quoted in ibid.

35. Quoted in Graves, "Hindu Nationalists Want Foreign Goods out of India."

36. Quoted in ibid.

37. See, for instance, Greer and Singh, *TNCs and India.*

38. Uniyal, "India—Economy."

39. Ibid.

40. Quoted in Graves, "Hindu Nationalists Want Foreign Goods out of India."

41. Lahiri, "Is Our Karma, Cola?"

42. Herd, "Burning Issue."

43. Pratap, "Ms. Greece Was Crowned 'Miss World' Despite the Opposition."

44. Ibid.

45. Ibid.

46. Jain, "Beauty Business."

47. Srinath, "Indian Women Bask in Glory as Rai Wins Miss World Title."

48. Quoted in Bavadam et al., "Beauty and the East," 57.

49. Ibid., 60.

50. Ibid., 59.

51. "Pageant Protests."

52. Singh, "Consumerism Is Not Development."

53. Lateef, "India's Rise to Globalism."

54. Jain, "Beauty Business."

55. Ibid.

56. Goldenberg, "Final Threat to India's Miss World."

57. Quoted in "Pageant Protests."

58. Stackhouse, "Polls Spotlight Cynicism, Doubt."

59. Quoted in Singh, "Consumerism Is Not Development."

60. Quoted in Goldenberg, "Final Threat to India's Miss World."

61. "Miss World Given Go-Ahead."

62. Quoted in Goldenberg, "Final Threat to India's Miss World."

63. Quoted in "Indian Protestors Urged to Let Miss World Contest Pass Off Peacefully."

64. Varam, "Security Heavy for Miss World Crowning in India."

65. Chopra, "TV Wins the Elections."

66. Ibid.

67. Ibid.

68. Quoted in "More Questions from Congress to Vajpayee."

69. Ibid.

70. Ibid.

71. Quoted in Misra and Pande, "Voters Are Angry at Political Turmoil."

72. Butalia, "Gandhi Meet Pepsi."

73. Ibid.

74. Ibid.

75. Ibid.

76. Ibid.

77. Ibid.

78. Ibid.

79. Foucault, "What Is an Author?"

80. Caughie, *Theories of Authorship.*

81. Ibid.

82. Spivak, "Can the Subaltern Speak?" and Foucault, "What Is an Author?" 447.

83. Ibid., 448.

84. Ibid., 448.

85. Ibid., 442.

86. Ibid., 450.

Chapter 5: Nikki Tonight, *Gandhi Today*

1. Mayo, *Mother India.*

2. Quoted in Nandy, *Traditions, Tyranny and Utopias,* 8.

3. Ibid.

4. Bataille, *Erotism, Death and Sensuality,* 64.

5. Ibid., 63.

6. Row Kavi, personal interview.

7. Ibid.

8. Row Kavi, "My Statement Was Immature, Juvenile."

9. Tim McGirk, "Gandhi Gaffe Dims Murdoch's Star."

10. Row Kavi, personal interview.

11. Fernandez, "Second Indian Nikki Show Victim Strikes Back."

12. "Saif Assaults Scribe, Kin over Film Review."

13. For other reviews of *Bombay Dost*, see Pais, "A Light in the Closet," and Thomas, "Unshackled Magazine Gives India's Invisible Gays a Voice."

14. Row Kavi, personal interview.

15. Waugh, "Queer Bollywood."

16. Ibid., 291.

17. Row Kavi, personal interview. In an interview with *Bombay Dost*, Akshay Kumar discussed the song-and-dance sequence from *Main Khiladi Tu Anari* and acknowledged the fact that he had a gay fan-following for this film. Alongside the interview, *Bombay Dost* published a picture of Akshay Kumar wearing nothing but a small towel. Anirudhh Chawda, in "Cracks in the Tinsel Closet," also made similar references to the film and reported the incident between Row Kavi and Saif Ali Khan in *Trikone*, a United States–based magazine for the South Asian gay and lesbian community.

18. Singh, "Saif's a Gentleman."

19. Ibid.

20. Gandhi, personal interview.

21. Ibid.

22. Ibid.

23. Ibid.

24. Ibid.

25. Ibid.

26. Ibid.

27. The case was filed in the Court of the Chief Metropolitan Magistrate at Esplanade, Mumbai, on July 3, 1995. The case is referred to in the court documents as "Tushar Gandhi vs. Rupert Murdoch and others," case no. 93/S of 1995.

28. *Bare Act*, 257.

29. Ibid., 257.

30. Ibid., 258–60.

31. Row Kavi filed his preliminary objection to Tushar Gandhi's defamation case on July 19, 1995. The objection claimed that the rationale for seeking a dismissal of the charges against the accused was based on legal precedent in K. M. Mathew vs. State of Kerala—1992(1) SEC 217.

32. "Tushar Gandhi sues STAR TV."

33. Ibid.

34. *Bare Act*, 260–61.

35. "'Nikki Tonight,' Controversial Talk Show Suspended."

36. Ibid.

37. Gandhi, personal interview.

38. Ibid.

39. Ibid.

40. Ibid.

41. Quoted in McGirk, "Gandhi Gaffe Dims Murdoch's Star."

42. Gandhi, "Politicians Have Turned Gandhiji into a Cheap Publicity Apparatus."

43. Gandhi, personal interview.

44. *Bare Act.*

45. "Defamation Warrant against Murdoch."

46. Ibid.

47. McGirk, "Gandhi Gaffe Dims Murdoch's Star."

48. Gandhi, personal interview.

49. "Bailable Warrant Issued to Murdoch in Chat Show."

50. Ibid.

51. Quoted in "Murdoch Wants STAR TV Out of Hong Kong before Handover."

52. Mullick and Mendonca, "STAR TV's Plans for Indian Base Gaining Ground?"

53. Ibid.

54. Quoted in "Murdoch Wants STAR TV Out of Hong Kong Before Handover."

55. Quoted in "Karnataka Govt. Denies Plan for STAR TV Venture in State."

56. Ibid.

57. Gandhi, personal interview.

58. Ibid.

59. Ibid.

60. Row Kavi, personal interview.

61. Ibid.

62. Similar arguments have been advanced by industry analysts who point to clear threats posed by transnational corporations to the traditional dominance of the big business houses in India. In an opinion piece published in the *Business Standard*, T. N. Ninan writes: "The noise about foreign domination is made by the country's 200 most influential families, who have enjoyed government protection all these years. They want the Indian consumer to be their captive"; "Weekend Ruminations." See also Aiyar, "Banias Who Run STAR TV."

63. Gandhi, personal interview.

64. Ibid.

65. Quoted in Bedi, "'Lost' Ashes of Gandhi Scattered in Ganges," 14.

66. "We Were Snubbed, Says Scion."

67. Gandhi, personal interview.

68. Row Kavi, personal interview.

69. Ibid.

70. Thomas, "Unshackled Magazine Gives India's Invisible Gays a Voice."

71. Demara, "Different Gay Dilemma in South Asia."

72. Ibid.

73. Ibid. See also Row Kavi, "Guiltless Homosexual Tradition?"

74. De, "Gay Missionary," 18.

75. Ibid., 18.

76. Ibid.

77. Wanvari, "'Bastard' Slight Brings Star's India Service into Disrepute."

78. Kanga, "Chat on a Hot Tin Roof."

79. Balram, "Glocal Girl."

80. Ibid.
81. Ibid.
82. McGirk, "Gandhi Gaffe Dims Murdoch's Star."
83. Balram, "Glocal Girl."
84. Ibid.
85. Ibid.
86. Appadurai, "Disjuncture and Difference in the Global Cultural Economy."
87. Featherstone, introduction to *Global Culture*, 6–7.
88. Balram, "Glocal Girl."
89. Hall, "The Local and the Global."
90. Ibid., 27.
91. Gandhi, personal interview.
92. Ibid.
93. "Tushar Gandhi Sues STAR TV."
94. "Ban STAR TV, Demand Elders."
95. "Gandhi-Bashing on Star TV Sparks Furore," 1.
96. "Ban STAR TV, Demand Elders."
97. Ibid.
98. Anderson, *Imagined Communities.*
99. See Cohen, "Holi in Banaras and the *Mahaland* of Modernity." Commenting on the intricacies of homosexual and homosocial relationships in India, he writes: "Intimately, among family and close friends of one's generation, sex is play, or *khel.* It is about joking around and about friendship, *dosti.* Different men may articulate the boundaries of friendship and play differently. For most, penetrative sex is seldom an idiom of play and in fact marks its boundary. The boys and (to a lesser extent) men who play around with friends their own age and of similar background must negotiate this mutual terrain of play" (417).
100. Row Kavi, personal interview.
101. Bataille, *Accursed Share.*
102. Ashraf, "Congress Disrupts Godse Play in Bombay."
103. Koppikar, "Mee Nathuram Boltoy."
104. Shankar, "Troubled Father Figure."
105. Ibid.
106. Sawant, "Making of the Mahatma."
107. Ibid.
108. Mauss, *Gift.*
109. Piel, "Bataille and the World," 99.

Conclusion

1. Pattanayak, "Problem," 12, emphasis added.
2. Ramanujan, "Where Mirrors Are Windows."
3. Ibid., 188.
4. Quoted in Ninan, *Through the Magic Window*, 5.
5. Ibid.
6. V. Jayanth, "Rekindling Patriotic Fervour."

7. Ibid.

8. Quoted in "The Good, Bad and Ugly."

9. Ibid.

10. Ibid.

11. Ibid.

12. Ninan, *Through the Magic Window*.

13. Indiresan, "Social and Economic Implications."

14. Ramanujan, "Is There an Indian Way of Thinking?" 42.

15. Kakar, *Inner World*, 15.

16. Shah, *Hype, Hypocrisy and Television in Urban India*, 246.

17. Ibid.

18. Ramanujan, "Is There an Indian Way of Thinking?" 42.

19. Rostow, *Stages of Economic Growth*.

20. See for instance, Ramesh, *Kautilya Today*, and Vittal, *Ideas for Action*.

21. Ramanujan, "Is There an Indian Way of Thinking?" 42.

22. Sonwalker, "India: Makings of Little Cultural/Media Imperialism?"

23. Ibid., 506.

24. Ramanujan, "Is There an Indian Way of Thinking?" 42.

25. Ram, "Modernist Anthropology and the Construction of Indian Identity."

26. Goldschmidt, foreword to Beals, *Village Life in South India*.

27. Tyler, *India: An Anthropological Perspective*.

28. Quoted in Ram, "Modernist Anthropology and the Construction of Indian Identity," 592.

29. Ibid., 593–94.

30. Unnikrishnan and Bajpai, *Impact of Television Advertising on Children*.

31. Ibid., 20.

32. Johnson, *Television and Social Change in Rural India*.

33. Ibid., 161–62.

34. Cited in Saksena, *Television in India*, 220.

35. Ramanujan, "Where Mirrors Are Windows."

36. Ibid., 189.

37. Ibid., 207.

38. Ibid., 207.

39. Ibid., 190.

40. Ramanujan, "Is There an Indian Way of Thinking?" 42–43.

41. Ibid.

42. Appadurai, "Here and Now."

43. Ibid.

44. Hall, "Signification, Representation, Ideology," 63.

45. Appadurai, "Here and Now."

46. Barker, *Television, Globalization and Cultural Identities*.

47. Spivak, "Can the Subaltern Speak?"

48. Nandy, *Intimate Enemy*.

49. Ibid.

BIBLIOGRAPHY

Agarwal, Amit. "MTV—Starting from Scratch: The Channel Seeks to Indianise Yet Retain Its Global Appeal." *India Today,* March 15, 1996, 160–61.

Aiyar, Swaminathan Anklesaria S. "The Banias Who Run STAR TV." *Times of India,* May 14, 1995, 14.

Althusser, Louis. *For Marx.* New York: Monthly Review, 1970.

Anand, M. "Sun TV: The Heat Is On." *Business World,* May 29, 2000, 20–26.

Anderson, Benedict. *Imagined Communities.* New York: Verso, 1991.

Appadurai, Arjun. "Disjuncture and Difference in the Global Cultural Economy." In *Global Culture,* edited by Mike Featherstone. London: Sage, 1990.

———. "Here and Now." In *Modernity at Large.* Minneapolis: University of Minnesota Press, 1997.

Ashraf, Syed Firdaus. "Congress Disrupts Godse Play in Bombay." Rediff.com India, July 17, 1998, available online at: www.rediff.com/news/1998/jul/17godse.htm.

Bahl, P. N. *Indira Gandhi: The Crucial Years (1973–1984).* New Delhi: Har-Anand, 1994.

"Bailable Warrant Issued to Murdoch in Chat Show." *Times of India,* July 11, 1996, 1.

Balram, Guvanthi. "The Glocal Girl." *Times of India,* April 9, 1995.

"Ban STAR TV, Demand Elders." *Times of India,* May 10, 1995.

Bare Act: Indian Penal Code, 1860. Allahabad: Law, 1996.

Bareth, Narayan. "Gandhi Fashion Image Row." BBC News, July 25, 2000, available online at: news.bbc.co.uk/1/hi/world/south_asia/850648.stm.

Barker, Chris. *Television, Globalization and Cultural Identities.* Buckingham, England: Open University Press, 1999.

Basu, Durga Das. *Introduction to the Constitution of India.* 18th ed. Nagpur: Wadhwa, 1999.

Bataille, Georges. *The Accursed Share: An Essay on General Economy.* 3 vols. New York: Zone Books, 1988–1991.

————. *Erotism, Death and Sensuality.* San Francisco: City Lights Books, 1986.

————. "The Notion of Expenditure." In *Visions of Excess: Selected Writings 1927–1939,* edited with an introduction by Alan Stoekl. Minneapolis: University of Minnesota Press, 1985.

Baudrillard, Jean. *Simulacra and Simulation.* Ann Arbor: University of Michigan Press, 1994.

Bavadam, Leela. "V Is for Variety," *Sunday* (India), January 1–7, 1995, 68–71.

Bavadam, Leela, Punam Thakur, Vasudha Gore, and Ghavdeep Kang. "Beauty and the East." *Sunday* (India), December 4–10, 1994, 52–61.

Bedi, Rahul. "'Lost' Ashes of Gandhi Scattered in Ganges." *Daily Telegraph,* January 31, 1997, 14.

Beltran, Luis Ramiro S. "Alien Premises, Objects and Methods in Latin American Communication Research." In *Communication and Development: Critical Perspectives,* edited by Everett M. Rogers. Beverly Hills, Calif.: Sage, 1977, 15–42.

Bhabha, Homi K. "DissemiNation: Time, Narrative and the Margins of the Nation." In *Nation and Narration,* edited by Homi Bhabha. New York: Routledge, 1991.

————. *The Location of Culture.* London: Routledge, 1994.

————. "Preface: Arrivals and Departures." In *Home, Exile Homeland,* edited by Hamid Naficy. London: Routledge, 1999.

Bhatia, Sidharth. "Cries of Cultural Imperialism Examined." *India Abroad,* May 26, 1995, 2.

Bhatt, S. C. *Satellite Invasion of India.* New Delhi: Gyan, 1994.

Butalia, Urvashi. "Gandhi Meet Pepsi." *Independent on Sunday* (London), April, 19, 1994, 19.

Campbell, Kurt C., and Thomas G. Weiss. "The Third World in the Wake of Eastern Europe." *Washington Quarterly* 14 (spring 1991): 91. Available online at: web. lexis-nexis.com.

Caughie, John, ed. *Theories of Authorship.* London: Routledge, 1981.

Chakrabarty, Dipesh. *Provincializing Europe: Postcolonial Thought and Historical Difference.* Princeton, N.J.: Princeton University Press, 2000.

"Channel V Listens to a New Song." *Cable and Satellite Asia,* January 1997, 18.

Chatterjee, Partha. *Nationalist Thought and the Colonial World: A Derivative Discourse?* London: Zed Books, 1986.

————. *The Nation and Its Fragments: Colonial and Postcolonial Histories.* Princeton N.J.: Princeton University Press, 1993.

Chatterji, P. C. *Broadcasting in India.* New Delhi: Sage, 1987.

Chawda, Aniruddh. "Cracks in the Tinsel Closet." *Trikone,* January 1996, 25–29.

Chopra, Mannika. "TV Wins the Elections." *Columbia Journalism Review* 37, no. 1 (1998): 67–69.

Cohen, Lawrence. "Holi in Banaras and the *Mahaland* of Modernity," *GLQ* 2, no. 4 (1995): 399–424.

"Cola Companies Battle On." *Economist,* November 8, 1989, available online at: web. lexis-nexis.com.

"Color Television Industry." India Infoline, July 28, 2000, available online at: www. indiainfoline.com.

Communicating by Satellite: Report of the Twentieth Century Fund Task Force on In-

ternational Satellite Communications. New York: Twentieth Century Fund Task Force, 1969.

Das, Sibabrata. "VSNL Set to Join Intersputnik Bandwagon." *Financial Express,* May 24, 1999, available online at: www.expressindia.com/fe/daily/19990524/fco24009p.htm.

De, Shobha. "The Gay Missionary." *Week* (India), November 6, 1994.

"Defamation Warrant against Murdoch." *Times of India,* July 5, 1995, 1.

Demara, Bruce. "A Different Gay Dilemma in South Asia: The Biggest Taboo Is Being Unmarried." *Toronto Star,* March 26, 1993, A25.

Derrida, Jacques. "Differance." In *Speech and Phenomena.* Evanston, Ill.: Northwestern University, 1973.

———. *Dissemination.* Translated by Barbara Johnson. Chicago: University of Chicago Press, 1981.

———. *Specters of Marx: The State of the Debt, the Work of Mourning, and the New International.* Translated by Peggy Kamuf, with an introduction by Bernd Magnus and Stephen Cullenberg. New York: Routledge, 1994.

Dhar, P. N. *Indira Gandhi, the "Emergency," and Indian Democracy.* New Delhi: Oxford University Press, 2000.

Dickenson, Nicole. 1996. "The Explosion of Asian Satellite TV." Reuters Textline, June 7, 1996, available online at: web.lexis-nexis.com.

"Direct to Home Broadcasting Sees Light of Day." *Connect,* February 20, 1999, available online at: www.connectmagazine.com/FEBRUARY%201999/Feb99/html/DTH.htm.

Donetsky, P. "Despite Major Difficulties a World Television System Is Possible." *Word,* December 1965, 51–52.

"Doordarshan Today." Doordarshan, available online at: ddindia.net, www.ddindia.net/Real/Content/about/2.html.

Drèze, Jean, and Amartya Sen. *India: Economic Development and Social Opportunity.* New Delhi: Oxford University Press, 1995.

Duncan, Emma. "Country Report: India." *Economist,* January 21, 1995, 3; available online at: web.lexis-nexis.com.

Esteva, Gustavo. "Development." In *The Development Dictionary: A Guide to Knowledge as Power,* edited by Wolfgang Sachs. London: Zed Books, 1995.

Etinger, Y. *NAM: History and Reality.* New Delhi: Allied, 1987.

Farrell, Stephen. "Now Sir Humphrey Gets to Say 'Yes Minister' in Hindi." *Times* (London), April 27, 2001, available online at: web.lexis-nexis.com.

"Fast Forward for a Slice of History." *Hindu,* November 13, 2001, available online at: web.lexis-nexis.com.

Featherstone, Mike. Introduction to *Global Culture,* edited by Mike Featherstone. London: Sage, 1990.

Fernandez, Clarence. "India: Reforms Trigger Clash with Indian Values." Reuter Textline, Reuter News Service, June 22, 1995, available online at: web.lexis-nexis.com.

———. "Second Indian Nikki Show Victim Strikes Back." Reuter News Service—India, May 10, 1995, available online at: web.lexis-nexis.com.

Ferraro, Nirmala. "Popular Reaction to Bombay TV." *Word,* December 1974, 30–36.

Foucault, Michel. "What Is an Author?" In *Language, Counter-Memory, Practice: Selected Essays and Interviews by Michel Foucault.* Ithaca, N.Y.: Cornell University Press, 1977.

Frenkel, Herbert M., and Richard E. Frenkel. *World Peace via Satellite Communications.* New York: Telecommunications Research Associates, 1965.

Fukuyama, Francis. "The End of History?" *National Interest* 16 (summer 1989): 3–18.

———. *The End of History and the Last Man.* New York: Free Press, 1992.

Galloway, Jonathan F. *The Politics and Technology of Satellite Communications.* Lexington, Mass.: Lexington Books, 1972.

"Gandhi-Bashing on Star TV Sparks Furore." *Times of India,* May 7, 1995, 1–2.

Gandhi, Tushar. "The Government Is Attracted by Murdoch's Lolly." *Times of India,* June 30, 1996.

———. Personal interview, August 2, 1996, Mumbai.

———. "Politicians Have Turned Gandhiji into a Cheap Publicity Apparatus." An interview with Joya Rajadhyaksha. *Times of India,* September 3, 1995.

Gargan, Edward. "New Delhi Journal: TV Comes In on a Dish, and India Gobbles It Up." *New York Times,* October 29, 1991, A4.

Gerbner, George, Hamid Mowlana, and Kaarle Nordenstreng, eds. *The Global Media Debate: Its Rise, Fall and Renewal.* Norwood, N.J.: Ablex, 1993.

Ghosh, Aparisim. "Marketing a Megabrand." *Business World,* November 3–16, 1993, 14–21.

Goldenberg, S. "Final Threat to India's Miss World," *South China Morning Post,* November 20, 1996, 19.

Goldschmidt, Walter. Foreword to *Village Life in South India,* by Alan R. Beals. Chicago: Aldine, 1974.

"The Good, Bad and Ugly." *Sunday Times of India,* June 18, 1995, 7.

Gopal, Sarvepalli. *The Anatomy of a Confrontation: The Babri Masjid-Ram Janmabhoomi Issue.* New Delhi: Viking Books, 1991.

Gramsci, Antonio. *Selections from the Prison Notebooks.* Hyderabad: Orient Longman, 1996.

Graves, Nelson. "Hindu Nationalists Want Foreign Goods Out of India." Reuters North American Wire, August 9, 1995, available online at: web.lexis-nexis.com.

Greer, Jed, and Kavaljit Singh. *TNCs and India: An Activist's Guide to Research and Campaign on Transnational Corporations.* New Delhi: Public Interest Research Group, 1996.

Guhathakurta, Subhrajit. "Electronics Policy and the Television Manufacturing Industry: Lessons from India's Liberalization Efforts." *Economic Development and Cultural Change* 42 (1994): 845–868.

Gupta, Nilanjana. *Switching Channels: Ideologies of Television in India.* Delhi: Oxford University Press, 1998.

Halbfass, Wilhelm. *India and Europe: An Essay in Understanding.* Albany, N.Y.: State University of New York Press, 1988.

Hall, Stuart. "Cultural Studies: Two Paradigms." *Media, Culture and Society* 2 (1980): 57–72.

———. "The Local and the Global: Globalization and Ethnicity." In *Culture, Globalization and the World-System,* edited by Anthony D. King. Minneapolis: University of Minnesota Press, 1997.

Halle, Louis J. "On Teaching International Relations." *Virginia Quarterly Review* 40, no. 1 (winter 1964): 11–25.

Herd, Deborah. "Burning Issue." *South China Morning Post,* December 4, 1996, TV Eye sec., 24.

Hickenlooper, Bourke B. "U.S. Sen. Hickenlooper Cites Facts in Support of Global Television Commission. *Television Opportunities,* July–August 1953, 2.

"How Cable TV Began and Spread in India." Ambez Media and Market Research, 1999, available online at: www.indiancabletv.net/cabletvhistory.htm.

" 'I Do Not Have Faith in Prasar Bharati': Minister Declares." Ipan Online, June 1999, available online at: ipan.com. www.ipan.com/reviews/archives/0699tv.htm

Illich, Ivan. *Medical Nemesis: The Expropriation of Health.* New York: Bantam Books, 1977.

"Indian Newcomers Find the Going Tough." Reuters Textline, April 1, 1996, available online at: web.lexis-nexis.com.

"Indian Protestors Urged to Let Miss World Contest Pass Off Peacefully." Agence France Press, November 4, 1996, available online at: web.lexis-nexis.com.

"India Seeking Votes in Bapu's Name." *Hindu,* February 9, 1998, available online at: web.lexis-nexis.com.

Indiresan, P. V. "Social and Economic Implications," *Seminar* 404 (May 1993): 14–23.

Innis, Harold. *Empire and Communication.* Oxford: Oxford University Press, 1950.

International Commission for the Study of Communication Problems. *Many Voices, One World: Communication and Society Today and Tomorrow: Towards a New More Just and More Efficient World Information and Communication Order.* Paris: UNESCO, 1980.

Jain, Madhu. "Hooked on War." *India Today,* February 15, 1991, 38–39.

Jain, Minu. "The Beauty Business," *Sunday* (India), December 4–10, 1994, 62.

Jalan, Bimal, ed. *The Indian Economy: Problems and Prospects.* New Delhi: Penguin Books, 1992.

Jayanth, V. "Rekindling Patriotic Fervour." *Hindu,* November 1, 1998, available online at: web.lexis-nexis.com.

Jeffrey, Robin. *India's Newspaper Revolution: Capitalism, Politics and the Indian-Language Press 1977–99.* New York: Oxford University Press, 2000.

Jha, Vinay. "Prasar Bharati Fate Hangs in Balance." *Statesman,* August 23, 1999, available online at: web.lexis-nexis.com.

Johnson, Kirk. *Television and Social Change in Rural India.* New Delhi: Sage, 2000.

Joshi, P. G. *Implications of India/U.S. Satellite Instructional Television Experiment.* Mumbai: Jai-Bharat, 2000.

Joshi, Vijay, and I. M. D. Little. *India's Economic Reforms: 1991–2001.* New Delhi: Oxford University Press, 1996.

Kakar, Sudhir. *The Inner World: A Psychoanalytic Study of Childhood and Society in India.* New Delhi: Oxford University Press, 1981.

Kanga, Fareeda. "Chat on a Hot Tin Roof." *Times of India,* April 4, 1995.

"Karnataka Govt. Denies Plan for STAR TV Venture in State," *Business Standard,* June 27, 1996, Hyderabad ed., 8.

Kellner, Douglas. *The Persian Gulf TV War.* Boulder, Colo.: Westview Press, 1992.

Khosla, Mukesh. "Channel Surfing." *Statesman,* May 26, 2001, available online at: web.lexis-nexis.com.

Koppikar, Smruti. "Mee Nathuram Boltoy: Hype and Hysteria." *India Today,* July 27, 1998, 77.

Kriplani, Manjeet. "The South Asian Invasion?" *Business Week,* July 5, 2001, available online at: web.lexis-nexis.com.

Kristof, Nicholas. "Satellites Bring Information Revolution to China." *New York Times,* April 11, 1993, sec. 1, 1.

Kumar, Keval J. *Mass Communication in India.* Mumbai: Jaico Press, 1999.

Kumar, Narendra, and Jai Chandiram. *Educational Television in India.* New Delhi: Arya Book Depot, 1967.

Kumar, Prem. *Television Industry in India: Market, Structure, Conduct and Performance.* New Delhi: Deep and Deep, 1988.

Kumar, Shanti. Review essay on *Screening Culture, Viewing Politics,* by Purnima Mankekar (1999) and *Politics after Television,* by Arvind Rajagopal (2000). *Screen* 44, no. 1 (spring 2003): 147–52.

Kumar, Shanti, and Michael Curtin. "'Made in India': In Between Patriarcy and Music Television," *Television and New Media* 3, no. 4 (2002): 345–66.

Laclau, Ernesto. "Hegemony and the Future of Democracy: Ernesto Laclau's Political Philosophy." In *Race, Rhetoric and the Postcolonial,* edited by Gary Olson and Lynn Worsham. Albany, N.Y.: State University of New York Press, 1999.

Lahiri, Indrajit. "Eenadu Chief Claims New Channel Has Penetrated All 1.3 M Cable Homes in A.P." *Times of India,* October 13, 1995.

Lahiri, Jaideep. "Is Our Karma, Cola?" *Economic Times,* May 27, 1996, 5.

Lateef, N. V. "India's Rise to Globalism," *Christian Science Monitor,* August 15, 1997, 20.

Lenin, Vladimir Il'ich. *Imperialism: The Last Stage of Capitalism.* Moscow: Progressive, 1966.

Lerner, Daniel. *The Passing of Traditional Society: Modernizing the Middle East.* New York: Free Press, 1958.

———. *Sykewar: Psychological Warfare against Germany, D-day to VE Day.* New York: Stewart, 1949.

Levey, Lawrence. "Democracy and Freedom Can Be Promoted via Global TV." *Television Opportunities* 5, no. 6 (July–August 1953): 1.

Luthra, H. R. *Indian Broadcasting.* Delhi: Publications Division, Ministry of Information and Broadcasting, Government of India, 1986.

Mankekar, Purnima. *Screening Culture, Viewing Politics: An Ethnography of Television, Womanhood, and Nation in Postcolonial India.* Durham, N.C.: Duke University Press, 1999.

Masani, Mehra. *Communication and Rural Progress.* Bombay: Leslie Sawhny Programme of Training in Democracy, 1975.

Mauss, Marcel. *The Gift.* Translated by Ian Cunnison. London: Routledge and Kegan Paul, 1966.

Mayo, Catherine. *Mother India.* New York: Harcourt, Brace, 1927.

McGirk, Tim. "Gandhi Gaffe Dims Murdoch's Star." *Independent* (London), July 6, 1995, 14.

McLuhan, Marshall. *The Gutenberg Galaxy: The Making of Typographic Man.* Toronto: University of Toronto Press, 1962.

Menon, Amarnath K. "Mahatma Brings Alive the Early Days and Little Known Facts of Gandhi as a Child." *India Today,* November 8, 1999, Television sec., 83.

Ministry of Information and Broadcasting. *An Indian Personality for Television: Report of the Working Group on Software for Doordarshan.* New Delhi: Government of India, 1982.

———. *Major Recommendations of the Working Group on Autonomy for Akashvani and Doordarshan.* New Delhi: Government of India, 1978, available online at: india.Indiantelevision.com/indianbroadcast/legalreso/legalresources.htm#.

———. *The Nitish Sengupta Committee Report on Prasar Bharati.* New Delhi: Government of India, 1996, available online at: india.Indiantelevision.com/ indianbroadcast/legalreso/legalresources.htm#.

———. *Radio and Television: Report of the Committee on Broadcasting and Information Media.* New Delhi: Government of India, 1965.

Misra, Neelesh, and Tanni Pande. "Voters Are Angry at Political Turmoil." *India Abroad,* April 18, 1997, 4.

"Miss World Given Go-Ahead," Reuters Textline, November 19, 1996, available online at: web.lexis-nexis.com.

Mittal, Renu. "Cable TV 'CAS'T Away." *Deccan Chronicle,* June 15, 2003, Sunday ed., 1.

Mitra, Ananda. *Television in India: A Study of the Mahabharata.* Thousand Oaks: Sage, 1993.

Mohapatra, P. K. "Film Music Weightage Is Going Down." *Financial Express,* December 9, 1999, available online at: www.financialexpress.com.

"More Questions from Congress to Vajpayee." *Hindu,* January 10, 1998, available online at: web.lexis-nexis.com.

Morley, David. *Television, Audiences and Cultural Studies.* New York: Routledge, 1993.

Mullick, Ashis, and Allen J. Mendonca. "STAR TV's Plans for Indian Base Gaining Ground?" *Economic Standard,* Hyderabad ed., June 26, 1996.

Mumford, Lewis. *Technics and Civilization.* New York: Harcourt Brace, 1993.

Mundt, Karl E. "We Can Give the World a Vision of America." *Baltimore Sun,* November 19, 1950.

"Murdoch Wants STAR TV Out of Hong Kong before Handover." Agence France Presse, June 21, 1996, finance pages sec.

Musolf, Lloyd D. *Communication Satellites in Political Orbit.* San Francisco: Chandler, 1986.

Nair, Unni R. "Thumpingly Successful Viji Thampi." Available online at: www. indiainfo.com/malayalam/interviews.

Nandy, Ashis. *The Intimate Enemy: Loss and Recovery of Self under Colonialism.* New Delhi: Oxford University Press, 1983.

———. "The Politics of Secularism and the Recovery of Religious Tolerance." *Alternatives* 13 (1988): 177–94.

———. *Traditions, Tyranny and Utopias: Essays in the Politics of Awareness.* New Delhi: Oxford University Press, 1987.

Nandy, Ashis, Shikha Trivedy, Shail Mayaram, and Achyut Yagnik. *Creating a Nation-*

ality: The Ramjanmabhumi Movement and Fear of the Self. New Delhi: Oxford University Press, 1995.

Nandy, Ashis, and Shiv Visvanath. "Modern Medicine and Its Non-Modern Critics: A Study in Discourse." In *Dominating Knowledge,* edited by Frederique Apffel Marglin and Stephen Apffel Marglin. Oxford: Oxford University Press, 1989.

Nasta, Dipak. "Trouble at Bombay Doordarshan." *India West,* June 5, 1996, 74.

Nehru, Jawaharlal. *Toward Freedom: The Autobiography of Jawaharlal Nehru.* New York: Day, 1941.

"'Nikki Tonight,' Controversial Talk Show Suspended." *Times of India,* May 12, 1995, 1.

Ninan, Sevanti. "Channel after Channel," *Hindu,* June 18, 2000, available online at: web.lexis-nexis.com.

———. "History of Indian Broadcasting Reform." In *Broadcasting Reform in India: Media Law from a Global Perspective,* edited by Monroe E. Price and Stefaan G. Verhulst. Delhi: Oxford University Press, 1998.

———. "Indelible Images." *Hindu,* October 7, 2001, available online at: web.lexis-nexis.com.

———. *Through the Magic Window.* New Delhi: Penguin Books, 1995.

Ninan, T. N. "Weekend Ruminations." *Business Standard,* October 25, 1997, available online at: http://1997.business-standard.com/97oct25/opinion1.html.

"One Crown, 1,500 Arrests at Miss World Pageant," Reuters, *Chicago Tribune,* November 24, 1996, available online at: web.lexis-nexis.com.

"Pageant Protests." *Asia Week,* December 6, 1996, Editorials sec., 15.

Pagedar, Pramod. "The Non-Aligned News Pool—A Distant Hope." *Word* (1979): 53–59.

Pais, Arthur J. "A Light in the Closet: India's Only Gay Publication." *Columbia Journalism Review* 31, no. 4 (November/December 1992): 14.

Pattanayak, D. P. "The Problem." *Seminar* 387 (November 1991): 12–15.

Pelton, Joseph N. "Key Problems in Satellite Communications: Proliferation, Competition, and Planning in an Uncertain Era." In *Economic and Policy Problems in Satellite Communications,* edited by Joseph N. Pelton and Marcellus S. Snow. New York: Praeger, 1977.

Pendakur, Manjunath, and Jyotsna Kapur. "Think Globally, Program Locally: Privatization of Indian National Television." In *Democratizing Communication,* edited by Mashoed Bailie and Dwayne Winseck. Cresskill, N.J.: Hampton Press, 1997.

Peters, John Durham. "Seeing Bifocally: Media, Place, Culture." In *Culture, Power, Place: Explorations in Critical Anthropology,* edited by Akhil Gupta and James Ferguson. Durham, N.C.: Duke University Press, 1997.

Piel, Jean. "Bataille and the World: From 'The Notion of Expenditure' to the Accursed Share." In *On Bataille,* edited by Leslie Anne Boldt-Irons. Albany, N.Y.: State University of New York Press, 1995.

Povaiah, Roshun. "A Fresh Focus." *Advertising and Marketing,* September 30, 1999, 36–41.

Prakash, Gyan. "Who Is Afraid of Postcoloniality?" *Social Text* 14, no. 4 (1996): 187–203.

Prasad, Bimal. "The Evolution of Non-Alignment." In *Issues before Non-Alignment: Past and Future,* edited by Uma Vasudev. New Delhi: Indian Council of World Affairs, 1983.

Prasad, Shishir. "Giving the Devil Its Due." *Business Standard,* August 19–25, 1997, available online at: http://1997.business-standard.com/97aug19/strategy/story1.htm.

"Prasar Bharati Bill." In *Government Media Autonomy and After,* edited by G. S. Bhargava. New Delhi: Concept, 1991.

Pratap, Anita. "Ms. Greece Was Crowned 'Miss World' Despite the Opposition." *CNN World View,* 6:04 A.M. ET (U.S.), transcript no. 96112306V36, Cable News Network, November 23, 1996, available online at: web.lexis-nexis.com.

Radhakrishnan, Radha. "Let's See." *Hindu,* November 16, 1995, available online at: web.lexis-nexis.com.

Rajadhyaksha, Ashish. "Beaming Messages to the Nation." *Journal of Arts and Ideas* 19 (May 1990): 33–52.

Rajagopal, Arvind. *Politics after Television: Hindu Nationalism and the Reshaping of the Public in India.* Cambridge: Cambridge University Press, 2001.

———. "The Rise of National Programming: The Case of Indian Television." *Media, Culture and Society* 15 (1993): 91–111.

Ram, Kalpana. "Modernist Anthropology and the Construction of Indian Identity." *Meanjin* 51, no. 3 (1992): 589–614.

Ramanujan, A. K. "Is There an Indian Way of Thinking?" In *India through Hindu Categories,* edited by McKim Marriot. Delhi: Sage, 1990.

———. "Where Mirrors Are Windows: Toward an Anthology of Reflections," *History of Religions* 28, no. 3 (1989): 186–224.

Ramesh, Jairam. *Kautilya Today: Jairam Ramesh on a Globalizing India.* New Delhi: India Research Press, 2002.

Rao, Lakshmana. "Communication and Development: A Study of Two Indian Villages." Ph.D. diss., University of Minnesota, 1963.

Rao, V. V. Personal interview, Hyderabad, India, July 25, 1996.

Rath, Claus-Dieter. "The Invisible Network: Television as an Institution in Everyday Life." In *Television in Transition,* edited by Philip Drummond and Richard Paterson. London: British Film Institute, 1985.

Razzaque, A. "Color Television Development at CEERI." *Electronics for You,* November 1982, 67–69.

Rettie, John. "India: Slow Stirrings of a Million Mutinies." Reuter Textline, *Guardian,* February 25, 1995, available online at: web.lexis-nexis.com.

Rist, Gilbert. *The History of Development: From Western Origins to Global Faith.* London: Zed Books, 1997.

Rogers, Everette. *Communication and Development: Critical Perspectives.* Beverly Hills, Calif.: Sage, 1977.

———. *Diffusion of Innovations.* New York: Free Press of Glencoe, 1962.

Rostow, W. W. *The Stages of Economic Growth: A Non-Communist Manifesto.* Cambridge: Cambridge University Press, 1960.

Row Kavi, Ashok. "A Guiltless Homosexual Tradition? How Hindu Religion and Society View Homosexuality." *Trikone* 11, no. 3 (July 1996): 8–9.

————. "My Statement Was Immature, Juvenile." Interview with S. Balakrishnan, *Sunday Times of India*, May 14, 1995, 10.

Sahai, Priya. "A Whiff of Promise." *Advertising and Marketing*, March 1–15, 1998, 37–51.

————. Personal interview, August 3, 1996, Mumbai.

"Saif Assaults Scribe, Kin over Film Review." *Times of India*, May 10, 1995, 6.

Saksena, Gopal. *Television in India: Changes and Challenges*. New Delhi: Vikas, 1996.

Samarajiva, Rohan. 1987. "The Murky Beginnings of the Communication and Development Field: Voice of America and 'The Passing of Traditional Society.'" In *Rethinking Development Communication*, edited by Neville Jayaweera and Sarath Amunugama. Singapore: Asian Mass Communication Research and Information Center.

Sarabhai, Vikram. *Science Policy and National Development*. Edited by Kamla Chowdhry. Delhi: Macmillan, 1974.

Sawant, Anagha. "Making of the Mahatma." *Indian Express*, July 15, 1998, available online at: www.expressindia.com.

Schramm, Wilbur, and Lyle Nelson. *Communications Satellite for Education and Development: The Case of India*. Washington, D.C.: Agency for International Development, 1968.

Schwoch, James. "'We Can Give the World a Vision of America': Crypto-Convergent Technology Approaches to 1950s Global TV." Unpublished paper presented at the Media and Cultural Studies Colloquium, Department of Communication Arts, University of Wisconsin-Madison, 2003.

Shah, Amrita. *Hype, Hypocrisy and Television in Urban India*. Delhi: Vikas, 1997.

Shankar, Lekha J. "A Troubled Father Figure." *Week*, January 18, 1999.

Shenon, Phillip. "A Race to Satisfy TV Appetites in Asia." *New York Times*, May 23, 1993, Sunday sec. 3, 12.

Shoesmith, Brian. "Footprints to the Future, Shadows of the Past: Toward a History of Communication Satellites in Asia." In *Beyond the Ionosphere: The Development of Satellite Communications*, edited by Andrew J. Butrica. Washington, D.C.: NASA Information Center, 1997, available online at: http://history.nasa.gov/SP-4217/ch17.htm.

Singh, Amrita. "Saif's a Gentleman." *Sunday Times of India*, May 14, 1995, 10.

Singh, Mayanika M. "The Muddle in the Middle." *Advertising and Marketing*, August 15, 1999, 118–20.

Singh, P. C. "Consumerism Is Not Development." *Indian Express*, February 13, 1997, 8.

Singhal, Arvind, and Everett M. Rogers. "The *Hum Log* Story in India." In *Entertainment-Education: A Communication Strategy for Social Change*. Mahwah, N.J.: Erlbaum, 1990.

Smith, Delbert D. *Communication via Satellite: A Vison in Retrospect*. Boston: Sijthoff, 1976.

Snow, Marcellus S. *International Commercial Satellite Communications: Economic and Political Issues of the First Decade of INTELSAT*. New York: Praeger, 1976.

Sonwalker, Prasun. "India: Makings of Little Cultural/Media Imperialism?" *Gazette* 63, no. 6 (2001): 505–19.

Spigel, Lynn. *Make Room for TV: Television and the Family Ideal in Postwar America.* Chicago: University of Chicago Press, 1992.

Spivak, Gayatri C. "Can the Subaltern Speak?" In *Marxism and the Interpretation of Culture*, edited by Cary Nelson and Lawrence Grossberg. Urbana: University of Illinois Press, 1988.

Srikanth, Sridevi. "BPL Plans to Set Up or Acquire Manufacturing Unit in UK." *Business Standard*, September 10, 1998, available online at: http://jul-sep98. business-standard.com/98sep10/corp15.htm.

Srinath, M. G. "Indian Women Bask in Glory as Rai Wins Miss World Title," *Deutsche-Presse Angentur*, November 20, 1994, available online at: web.lexis-nexis.com.

Srinivasan, Raman. "No Free Launch: Designing the Indian National Satellite." In *Beyond the Ionosphere: The Development of Satellite Communications*, edited by Andrew J. Butrica. Washington, D.C.: NASA Information Center, 1997, available online at: http://history.nasa.gov/SP-4217/ch16.htm.

Srinivasan, T. N. *Eight Lectures on India's Economic Reforms.* New Delhi: Oxford University Press, 2000.

Stackhouse, J. "Polls Spotlight Cynicism, Doubt." *Globe and Mail*, August 11, 1997, A8, available online at: web.lexis-nexis.com.

Swami, Praveen. "Autonomy in Prospect." *Frontline*, September 20–October 3, 1997, available online at: www.flonnet.com/fl1419/14191290.htm.

"Synergy of Sight and Sound." *Hindu*, March 1, 1997, 39.

Thomas, Christopher. "Unshackled Magazine Gives India's Invisible Gays a Voice." *Times* (London), November 27, 1991, available online at: web.lexis-nexis.com.

Thompson, John. 1995. *Media and Modernity.* Cambridge: Polity Press, 1995.

Tomlinson, John. *Cultural Imperialism.* Baltimore: Johns Hopkins University Press, 1991.

"Tushar Gandhi Sues STAR TV." *Times of India*, May 9, 1995, 6.

Tyler, Stephen A. *India: An Anthropological Perspective.* Pacific Palisades, Calif.: Goodyear, 1973.

UNESCO (United Nations Educational, Scientific and Cultural Organization). *Television: A World Survey.* Paris: UNESCO, 1953; reprint, New York: Arno Press, 1972.

Unger, Arthur. "TV Comes to India: A Talk with Its Top Broadcast Official." *Christian Science Monitor*, March 22, 1985, 25.

Uniyal, Mahesh. "India—Economy: TNCs Get the Stick from Angry Legislators." *Inter Press Service*, July 17, 1995, available online at: web.lexis-nexis.com.

Unnikrishnan, Namita, and Shailaja Bajpai. *The Impact of Television Advertising on Children.* New Delhi: Sage, 1996.

Varam, Dev. "Security Heavy for Miss World Crowning in India," Reuters Textline, November 23, 1996, available online at: web.lexis-nexis.com.

Varma, Sharad. "Air Waves: Public Property." *Lawyers* 10, no. 5 (May 1995): 4–11.

Vittal, N. *Ideas for Action.* Delhi: MacMillan. 2002.

Wallerstein, Immanuel. *The Capitalist World Economy.* Cambridge: Cambridge University Press, 1980.

Wanvari, Anil. "'Bastard' Slight Brings Star's India Service into Disrepute." *Asian Advertising and Marketing,* Broadcast sec., May 19, 1995, 4.

Wanwari, Anil. "The Pay TV Conundrum." Indian Television Dot Com, October 1996, available online at: www.indiantelevision.com/viewpoint/conundrum.htm.

Watson, Miranda. "Asia's Rising Star." *Cable and Satellite Express,* May 23, 1996, 11.

Waugh, Thomas. 2003. "Queer Bollywood, or I'm the Player, You're the Naive One: Patterns of Sexual Subversion in Recent Indian Popular Cinema." *In Keyframes: Popular Cinema and Cultural Studies,* edited by Matthew Tinkcom and Amy Villarejo. New York: Routledge, 2003.

Weidner, Edward W. *Technical Assistance in Public Administration Overseas.* Chicago: Public Administration Services, 1964.

"We Were Snubbed, Says Scion." *Courier Mail,* August 16, 1997, 34, available online at: web.lexis-nexis.com.

Williams, Raymond. *Television: Technology and Cultural Form.* New York: Schocken Books, 1975.

Zee TV, *Annual Report,* 1996.

INDEX

SHANTI KUMAR is an assistant professor in the Department of Communication Arts at the University of Wisconsin–Madison. He is the coeditor of *Planet TV: A Global Television Reader.* He has also published book chapters in edited anthologies and articles in journals such as *Television and New Media, Jump Cut,* the *Quarterly Review of Film and Video,* and *South Asian Popular Culture.* His research and teaching interests include television and cultural studies, new media technologies, global media studies, and postcolonial theory and criticism.

Popular Culture and Politics in Asia Pacific

The University of Illinois Press
is a founding member of the
Association of American University Presses.

———————————————————————

Composed in 10.5/13 Adobe Minion
with Minion and Caravan 2 display
by Celia Shapland
for the University of Illinois Press
Designed by Paula Newcomb
Manufactured by Thomson-Shore, Inc.

University of Illinois Press
1325 South Oak Street
Champaign, IL 61820-6903
www.press.uillinois.edu